Rodent Models of Neurodegenerative Disease

Rodent Models of Neurodegenerative Disease explores neurodegenerative disorders such as Alzheimer's disease, Parkinson's disease, Huntington's disease, amyotrophic lateral sclerosis, frontotemporal dementia, Lewy body disease, and prion disease which represent some of the most complex and devastating challenges in neuroscience. With rising prevalence in aging populations worldwide, these disorders place immense burdens on patients, families, and healthcare systems. This comprehensive volume, *Rodent Models of Neurodegenerative Disease*, provides an in-depth examination of the utility, strengths, and limitations of rodent models in unraveling the molecular, cellular, and behavioral aspects of neurodegeneration. Organized by disease, each chapter explores both genetic and non-genetic models, highlighting their relevance to human pathology, translational value, and contributions to therapeutic discovery.

Key features of this book include the following:

- Detailed coverage of rodent models for major neurodegenerative diseases.
- Critical analysis of disease mechanisms, from protein aggregation and mitochondrial dysfunction to neuroinflammation and synaptic loss.
- A dedicated chapter on behavioral assays, bridging molecular pathology with motor, cognitive, and social outcomes.
- Insights into the translational challenges of preclinical research and its role in shaping future therapies.

With contributions from leading neuroscientists, this book serves as a valuable reference for researchers, graduate students, and professionals in neuroscience, neurology, pharmacology, and related fields. Whether for experimental design, model selection, or therapeutic evaluation, it provides an essential guide for advancing neurodegenerative disease research.

About the Editors:

Dr. Pegah Javadpour, PhD, is Assistant Professor of Medical Physiology at the Neuroscience Research Center, Institute of Neuroscience and Cognition, Shahid Beheshti University of Medical Sciences, Tehran, Iran. Her research focuses on mitochondrial dysfunction, oxidative stress, and neuroinflammation in neurodegenerative disorders, particularly Parkinson's and Alzheimer's diseases. She has a special interest in mitochondria-targeted therapeutic strategies, including mitochondrial transplantation therapy (mitotherapy), as potential treatments for these conditions.

Dr. Rasoul Ghasemi, PhD, is Associate Professor of Medical Physiology in the Department of Physiology, Shahid Beheshti University of Medical Sciences, Tehran, Iran. His research explores the neurophysiology of learning and memory, with emphasis on insulin signaling, hippocampal function, neuroinflammation, oxidative stress, and neuroprotection in neurodegenerative diseases. He has published extensively on the cellular and molecular mechanisms underlying brain resilience and pathology.

Neuroscience of Neurodegenerative Disorders

Series Editor Wael Mohamed

Wael Mohamed Yousef Mohamed

BBCH (Egypt), MMSc (Egypt), MD/PhD (USA)

Dr. Mohamed earned his PhD from Penn State University (USA) and currently is an Associate Professor at IIUM (International Islamic University Malaysia) Medical School in Malaysia. He serves as an editor for *Frontiers in Neurology* and PLOS-ONE Journals. His research is well supported by extramural funding. He is the founder and chair of the AfrAbia-PD-Genomic Consortium.

About This Series

This book series aims to publish volumes focusing on both basic and clinical research in the field of Neurodegenerative Disorders with a focus on delineating the recent advances in various disorders including Parkinson's Disease, Alzheimer's Disease, ALS, Huntington's Disease, among other common Neurodegenerative disorders. The books will examine different disorders aspects including pathophysiology, diagnostic, treatment, and precision medicine. Volumes in the series will also highlight the recent advances in the management of Neurodegenerative disorders especially in OMICS and Genomic era and precision medicine. Volumes will be useful for neuroscientists, neurologists, psychiatrists, psychologists, neurosurgeons, and those interested in preventive medicine.

Integrating Neuroimaging, Computational Neuroscience, and Artificial Intelligence
Edited by Indranath Chatterjee and Nasrallah Moradikor

Non-Pharmacological Interventions for Dementia: A Comprehensive Approach
Edited by Abraham Olufemi Asuku, Ebele Joan Ajagun, and Gbonjubola Oyinlola Ogungbangbe

Rodent Models of Neurodegenerative Disease
Edited by Pegah Javadpour and Rasoul Ghasemi

Rodent Models of Neurodegenerative Disease

Edited by

Pegah Javadpour and Rasoul Ghasemi

CRC Press
Taylor & Francis Group
Boca Raton London New York

CRC Press is an imprint of the
Taylor & Francis Group, an **informa** business

Designed cover image: Shutterstock

First edition published 2026
by CRC Press
2385 NW Executive Center Drive, Suite 320, Boca Raton FL 33431

and by CRC Press
4 Park Square, Milton Park, Abingdon, Oxon, OX14 4RN

CRC Press is an imprint of Taylor & Francis Group, LLC

ISBN: 978-1-032-66049-3 (hbk)
ISBN: 978-1-032-67651-7 (pbk)
ISBN: 978-1-032-67652-4 (ebk)

DOI: 10.1201/9781032676524

Typeset in Times
by KnowledgeWorks Global Ltd.

Contents

Preface

Neurodegenerative diseases represent some of the most debilitating and complex challenges affecting the human brain. They carry profound consequences not only for patients and their families but also for healthcare systems worldwide. As our population ages, conditions such as Alzheimer's disease, Parkinson's disease, Huntington's disease, amyotrophic lateral sclerosis, and prion disease are becoming increasingly prevalent. This trend underscores the urgent need for innovative research and novel treatment strategies to address these devastating disorders.

This book, *Rodent Models of Neurodegenerative Disease*, emerges from a collective dedication to advancing our understanding of these intricate disorders. By examining the utility, strengths, and limitations of rodent models, we strive to bridge the gap between laboratory research and clinical applications. Rodents, particularly mice and rats, are invaluable tools in neuroscience, thanks to their genetic tractability, physiological similarities to humans, and capacity to replicate essential pathological features of neurodegenerative diseases.

The chapters in this volume provide a comprehensive overview of the most widely used rodent models, organized by disease. Each section delves into the genetic and non-genetic models that replicate the hallmarks of neurodegeneration, from protein aggregation and synaptic dysfunction to neuroinflammation and behavioral deficits. We highlight not only the technical aspects of these models but also their translational relevance, offering insights into how preclinical findings can inform human studies. Additionally, one dedicated chapter reviews the spectrum of rodent behavioral assays employed to assess different aspects of neurodegenerative diseases. These paradigms cover domains such as cognition, emotion, motor coordination, and social interaction, providing essential links between molecular pathology and functional outcomes.

This book is designed for researchers, clinicians, and students in neuroscience, neurology, and related fields. Whether you are new to the study of neurodegenerative diseases or an experienced investigator, we hope this work serves as a valuable resource to guide your research and inspire new discoveries.

Finally, we extend our deepest gratitude to the contributors whose expertise and dedication made this book possible. Their collective efforts reflect the collaborative spirit of scientific inquiry and the shared goal of unraveling the mysteries of neurodegenerative diseases.

The Editors

Contributors

Fatemeh Abbaszadeh
Neurobiology Research Center, Institute of Neuroscience and Cognition, Shahid Beheshti University of Medical Sciences, Tehran, Iran

Fariba Akhondzadeh
Department of Physiology, School of Medicine, Shahid Beheshti University of Medical Sciences, Tehran, Iran

Sahar Askari
Department of Physiology, School of Medicine, Iran University of Medical Sciences, Tehran, Iran

Saeid Bagheri-Mohammadi
Department of Paramedicine, Amol School of Paramedical Sciences, Mazandaran University of Medical Sciences, Sari, Iran

Bita Firouzan
Neuroscience Research Center, Institute of Neuroscience and Cognition, Shahid Beheshti University of Medical Sciences, Tehran, Iran
Department of Neuroscience, Northwestern University Feinberg School of Medicine, Chicago, Illinois, USA

Rasoul Ghasemi
Department of Physiology, School of Medicine, Shahid Beheshti University of Medical Sciences, Tehran, Iran
Neurophysiology Research Center, Institute of Neuroscience and Cognition, Shahid Beheshti University of Medical Sciences, Tehran, Iran

Pegah Javadpour
Neuroscience Research Center, Institute of Neuroscience and Cognition, Shahid Beheshti University of Medical Sciences, Tehran, Iran

Fatemeh Mokhtari Naseri
Department of Physiology, School of Medicine, Shahid Beheshti University of Medical Sciences, Tehran, Iran

1 Introduction to Animal Models in Neuro-degenerative Research

Bridging the Gap Between Lab and Clinic

Rasoul Ghasemi

The greatness of a nation and its moral progress can be judged by the way its animals are treated.

Mahatma Gandhi

1.1 INTRODUCTION

Since ancient times, the human brain—its functions and associated diseases—has been a subject of inquiry for investigators seeking to understand its complexities. The earliest known reference to brain function dates back to the 17th century BC, when the symptoms, diagnosis, and prognosis of patients with head injuries were described (1). Over time, various philosophers and scientists have endeavored to explore the nature of the brain and the consequences of its dysfunction.

In this regard, Aelius Galenus, known as Galen (130–200 CE), a prominent Roman-Greek physician, was among the first to study animal brains to gain insights into human brain function. Through dissections of various animals, including oxen, apes, swine, and even elephants, he concluded that the cerebellum, being denser, is responsible for muscle control, while the softer cerebrum is the center of sensory processing (2). These studies of animal brains were among the first known efforts to access the brain by using animal models. Since then, numerous studies on non-human brains have significantly advanced our understanding of brain anatomy, physiology, and pathology. Among these, research on neurodegenerative disorders has gained prominence more recently. These disorders are still not fully understood, particularly in terms of their pathophysiology and treatment strategies, highlighting the need for more comprehensive studies. To conduct such research, it is essential to develop animal models that closely mimic the human condition.

As mentioned previously, Galen conducted studies on the brain using various animals and discovered that the brains of apes exhibit the closest resemblance to the human brain. However, due to several factors, rodents, particularly rats and mice, are commonly employed in such studies and have advantages over primates. These make rodents the best and most frequent choice for such studies to the extent that, in one estimation, it is believed that 90% of animal studies are conducted on rodent models (3).

1.2 WHY RODENT MODELS

One primary reason rodent models are widely employed in studies of human physiology and pathology is the sequencing of the mouse genome, which reveals remarkable similarities to the human genome.

DOI: 10.1201/9781032676524-1

It is estimated that approximately 99% of the mouse genome has a human equivalent, enabling researchers to more effectively extrapolate experimental results to human conditions (4).

A key factor contributing to the preference for rodent models is their high reproducibility. Rats and mice exhibit a short estrous cycle and a gestation period of only about 20 days. Typically, a single litter consists of four to eight pups. Furthermore, these offspring mature rapidly, reaching reproductive age within just seven weeks. This fast reproductive cycle facilitates high breeding rates, making rodents an ideal choice for studies requiring large sample sizes. In addition, rodent breeding is not only easier but also significantly more cost-effective than that of larger animals. Their small size makes housing and environmental control relatively simple, and their dietary needs are inexpensive and easy to fulfill, factors that collectively enhance their practicality in laboratory settings.

Another advantage of rodents in biomedical research is their relatively small size, which simplifies handling and allows the use of smaller, less costly laboratory instruments for procedures such as surgery and behavioral testing. Furthermore, because the amount of drug administered in experimental studies is generally proportional to the animal's body weight, using smaller animals like rodents reduces the overall cost of research materials. Recent advances in genetic engineering have also made rodents, particularly mice, a cornerstone of experimental biology. The feasibility of genetic manipulation has led to the development of numerous inbred strains and transgenic models, further broadening their utility in studying complex human diseases.

Beyond their general advantages, rodents are particularly well-suited for modeling neurodegenerative disorders. These conditions are often age-related, and due to the relatively short lifespan of rodents, age-associated changes can be studied within a manageable time frame. This makes rodents especially valuable for investigating the progression and treatment of neurodegenerative diseases.

1.3 DISADVANTAGES OF RODENT MODELS

While the use of animals in experimental studies offers numerous advantages, it is important to acknowledge several drawbacks, particularly when using rodent models to study neurodegenerative disorders, as discussed in the preceding sections.

One key limitation is that rodents are often considered suboptimal models for studying inflammation and diseases with an inflammatory basis. A common argument against their use is their relative resistance to inflammation and infection, which differs from human responses (5). This concern is supported by multiple studies that have demonstrated limited success in translating findings from rodent models to clinical outcomes in human trials—especially in the context of inflammatory conditions. Given that neurodegenerative disorders are strongly associated with chronic inflammation, this limitation is especially relevant and problematic for modeling such diseases in rodents.

Another major drawback lies in the complexity of human diseases like neurodegenerative disorders, which are influenced by a lifetime of interacting genetic, environmental, and toxicological factors. Animal models cannot fully recapitulate this multifactorial nature.

For instance, in Alzheimer‘s disease, the progression of symptoms in humans results from a combination of aging and numerous internal and external factors. These human-specific dynamics are not accurately mirrored in rodents, despite apparent similarities in pathological features.

Additionally, many rodent experiments are conducted during the daytime under artificial lighting conditions. Since rodents are nocturnal, their physiological and behavioral responses can be significantly altered when studied outside of their natural activity cycle. This introduces a confounding variable, especially in behavioral studies of neurodegenerative disorders. The best way to prevent such distortion is possibly to test the animals during their active phase (night) (6). Given the fact that rodents' visual system is less sensitive to red and red/orange lights, some proposed housing and working with rodents under the sodium lamps, with red illumination, enabling the researchers to visualize and test the animals in the dark period of animals, and minimizing the effects of light on rodent behavior (7). This idea, however, is challenged because of the fact that rodents are not completely blind to red/orange lights. Therefore, these lights cannot totally mimic the darkness for animals (6). An alternative approach is the development of automated testing systems that enable continuous, undisturbed monitoring during their active phase (darkness). Furthermore, it has been recommended that studies should explicitly report the timing and lighting conditions under which tests are conducted to improve reproducibility and transparency (6).

While earlier sections highlighted the advantages of rodents, such as their small size, lower cost, ease of handling, and high reproductive rate, these same traits may also contribute to reduced ethical vigilance in their care. Compared to larger animals like primates, which are more expensive and slower to reproduce, rodents are sometimes subjected to less rigorous ethical scrutiny. This adds to the broader ethical debate surrounding the use of animals in medical research, including research on neurodegenerative diseases. These debates, which trace back to the 19th century, are grounded in the recognition that animals are sentient beings capable of experiencing pain and suffering.

There are two general positions in this debate: on one end are animal rights and protection groups who argue for the complete prohibition of animal use in research; on the other, some advocate for stricter regulations to minimize animal usage and suffering (8). The latter position supports a cost-benefit analysis for each study, weighing the potential scientific value against the ethical cost, and emphasizes the need to pursue alternative methods wherever possible (9). To enhance the translational validity of animal models, one proposed strategy is to begin with a comprehensive genomic and molecular profiling of the disease in humans. Researchers can then use this information to develop animal models that not only display similar phenotypes but also replicate the molecular pathways involved in human disease.

Ultimately, it is important to recognize that just as no single human can represent the entire human population, no animal model can fully replicate the complexity of human diseases. Although humans themselves would, in theory, serve as the most accurate models for studying human conditions, in practice, such research is fraught with ethical and logistical challenges. The following section will explore in greater detail the reasons why direct human experimentation remains largely infeasible.

1.4 WHY NOT HUMAN STUDIES

There is no doubt that the most valuable information regarding brain function and pathology could be obtained by studying the human brain. However, for several reasons, it is not always feasible or ethical to investigate brain pathology directly, except in clinical settings where patients serve as subjects. Even in such cases, the extent and location of brain damage caused by disease or injury cannot be controlled, and patients often have diverse medical histories. These factors make it difficult to standardize variables, and meaningful results typically require large sample sizes. In human studies, numerous uncontrollable variables exist across different populations. For example, it is nearly impossible to assemble a group of individuals with identical health-determining factors. Additionally, due to technical and ethical constraints, most invasive techniques cannot be performed on humans. While non-invasive approaches such as imaging methods are commonly used, they often lack the spatial and temporal resolution required to study intricate neuronal circuits in detail.

REFERENCES

1. Breasted J. The Edwin Smith Surgical Papyrus: published in facsimile and hieroglyphic transliteration with translation and commentary in two volumes. 1930 [cited 2025 Apr 9]; Available from: https://ixtheo.de/Record/1076082262
2. Freemon FR. Galen's ideas on neurological function. Journal of the History of the Neurosciences [Internet]. 1994 Oct 1 [cited 2025 Apr 9];3(4):263–71. Available from: https://www.tandfonline.com/doi/abs/10.1080/09647049409525619
3. Kiani AK, Pheby D, Henehan G, Brown R, Sieving P, Sykora P, et al. Ethical considerations regarding animal experimentation. Journal of Preventive Medicine and Hygiene [Internet]. 2022 Jun 1 [cited 2025 Apr 9];63(2 Suppl 3):E255. Available from: https://pmc.ncbi.nlm.nih.gov/articles/PMC9710398/

4. Li H, Auwerx J. Mouse systems genetics as a prelude to precision medicine. Trends in Genetics : TIG. 2020 Apr;36(4):259–72.
5. Breschi A, Gingeras TR, Guigó R. Comparative transcriptomics in human and mouse. Nature Reviews Genetics. 2017 Jul;18(7):425–40.
6. Hawkins P, Golledge HDR. The 9 to 5 Rodent – time for change? Scientific and animal welfare implications of circadian and light effects on laboratory mice and rats. Journal of Neuroscience Methods. 2018 Apr;300:20–5.
7. McLennan IS, Taylor-Jeffs J. The use of sodium lamps to brightly illuminate mouse houses during their dark phases. Laboratory Animals. 2004 Oct;38(4):384–92.
8. Hajar R. Animal testing and medicine. Heart Views : The Official Journal of the Gulf Heart Association. 2011 Jan;12(1):42.
9. Khoo SYS. Justifiability and animal research in health: can democratisation help resolve difficulties? Animals: An Open Access Journal from MDPI. 2018 Feb;8(2).

1 Rodent Models of Alzheimer's Disease

Pegah Javadpour

1.1 INTRODUCTION

Alzheimer's disease (AD) is a progressive neurodegenerative disorder characterized by cognitive decline, memory loss, and behavioral changes. It is the most common cause of dementia, affecting millions of individuals worldwide. The primary pathological hallmarks of AD include the accumulation of amyloid-beta (Aβ) plaques, tau neurofibrillary tangles (NFTs), neuroinflammation, and synaptic dysfunction, ultimately leading to neuronal loss. Despite extensive research, effective disease-modifying treatments remain elusive, necessitating a deeper understanding of disease mechanisms and the development of novel therapeutic approaches. In addition to being a significant medical challenge, AD also presents a growing socioeconomic burden. The prevalence of AD is steadily increasing due to aging populations worldwide, placing immense pressure on healthcare systems and caregivers. The economic impact of AD extends beyond medical costs to lost productivity and the emotional toll on families. Addressing these challenges requires robust preclinical models to study disease progression and evaluate potential therapeutic interventions.

This chapter aims to provide a comprehensive overview of rodent models used in the study of AD, highlighting their advantages, limitations, and translational relevance. It delves into specific AD models, including genetically engineered and neurotoxin-induced models, detailing their underlying pathophysiological mechanisms and the behavioral, biochemical, and histological assessments used to evaluate disease progression and therapeutic interventions.

In this chapter, the symptoms of AD, including cognitive, behavioral, and functional impairments observed in patients, will first be outlined. Following this, rodent models of AD will be examined, beginning with models induced by neurotoxins such as streptozotocin, aluminum, and Aβ peptides, which are widely used to mimic specific aspects of AD pathology. Subsequently, genetically engineered models will be explored. These models, including transgenic mice carrying mutations in amyloid precursor protein (APP), PSEN1, and PSEN2 genes, replicate key features of the disease, providing insight into genetic risk factors and molecular mechanisms underlying AD.

1.2 ALZHEIMER'S DISEASE NEUROPATHOLOGY

AD is characterized by two major components: (A) a clinical phenotype that manifests as progressive dementia affecting various cognitive abilities and (B) distinct neuropathological features, including the deposition of Aβ in both extracellular plaques and blood vessels, known as cerebral amyloid angiopathy (CAA). Additionally, neuritic plaques (NPs), a subset of amyloid plaques, are defined by the presence of hyperphosphorylated tau protein (p-tau), which also forms intraneuronal aggregates that lead to NFTs. These pathological features often correlate with early synaptic loss, neuronal loss, gliosis, mitochondrial dysfunction, neuroinflammation, gut microbiome alterations, brain atrophy, and disruptions in neurovascular function and blood–brain barrier (BBB) integrity.[1–5]

In AD, cognitive impairment initially presents as deficits in short-term memory. However, as the disease progresses, additional cognitive domains, including language, motor function, and attention, become progressively impaired.

DOI: 10.1201/9781032676524-2

The preclinical stage of AD, known as the "cellular phase," is characterized by early alterations in neurons, astrocytes, and microglia before cognitive symptoms manifest. During this phase, neuroinflammation, vascular dysfunction, aging, and glymphatic system impairment either drive or occur alongside the accumulation of Aβ, contributing to disease progression.[6] Aβ is thought to trigger the spread of tau pathology, although the exact mechanism remains unknown. The accumulation of tau pathology is related to the presence of necroptosis markers in neurons exhibiting granulo-vacuolar degeneration, a hallmark of neuronal stress and dysfunction.[6]

Although memory loss, language impairment, and cognitive decline are defining features of AD, disruptions in the sleep–wake cycle, leading to prolonged wakefulness at night and increased daytime sleepiness, are also common. Moreover, patients exhibit a significant reduction in slow-wave sleep (SWS) and rapid eye movement (REM) sleep, both of which are crucial for memory consolidation and cognitive function.[7,8]

Advanced age remains the strongest risk factor for AD. Genetically, the APOE ε4 allele is the most significant contributor to AD susceptibility. Additionally, women, especially those over 80 years old, have a higher likelihood of developing AD compared to men. Although both sexes exhibit a similar Aβ burden, women with AD tend to show greater tau pathology, which is strongly associated with disease progression.[6,9] While significant progress has been made in understanding disease mechanisms, further research is needed to elucidate the precise interactions between these pathological processes. Animal models, such as neurotoxin-induced models and transgenic mice, have proven invaluable in studying AD's progression and pathophysiology. These models allow for controlled investigations into the effects of amyloid and tau accumulation, as well as neuroinflammation and vascular changes, which closely mimic aspects of human AD.

1.3 RODENT MODELS OF ALZHEIMER'S DISEASE

1.3.1 Non-Genetic Models

1.3.1.1 Aluminum

Aluminum is recognized as a significant environmental factor contributing to neurodegenerative diseases, particularly AD. Early studies on experimental animals, notably those by Klatzo et al.[10] and Terry and Peña (1986), highlighted the neurotoxic effects of aluminum. These studies demonstrated that injecting aluminum salts into rabbit brains led to the formation of NFTs, similar to those observed in AD.[10] Subsequent research by Yumoto et al.[11], using an isotope tracer, revealed that aluminum injected subcutaneously into pregnant rats could cross the placental barrier and accumulate in fetal brain cells.[11]

Chronic daily subcutaneous injections of aluminum in young New Zealand White rabbits resulted in the accumulation of intranuclear aluminum within NFTs in specific brain regions. Initially, these accumulations were detected in the spinal cord and brain stem, with no significant deposits in the hippocampus. Interestingly, these rabbits did not exhibit neurological symptoms.[12]

Further investigations showed that the intracisternal administration of aluminum maltolate induced apoptosis through mitochondrial damage and endoplasmic reticulum (ER) stress in New Zealand White rabbits. These findings led researchers to propose aluminum as a compound for developing animal models of neurodegenerative diseases to facilitate mechanistic studies and assess potential neuroprotective compounds.[13]

Immunohistochemical studies on aged rabbit brains treated with aluminum maltolate revealed neuropathological features resembling AD, including NFTs, oxidative stress, paired helical filaments (PHF1), and Aβ deposition. These pathological events co-localized with apoptotic neuronal cells, particularly in the hippocampus.[14]

Additionally, activation of neuroinflammatory pathways such as NLR family pyrin domain containing 3 (NLRP3) and increased levels of pro-inflammatory cytokines like interleukin-1 (IL-1) have been observed in various brain regions following aluminum chloride-induced AD in Wistar rats.[15]

Furthermore, sub-chronic aluminum chloride exposure in rats, from birth until postnatal day 90, resulted in impaired learning and memory, structural changes in hippocampal synapses, and neuronal alterations.[16] Intrahippocampal administration of aluminum chloride in rats also led to recall deficits and a reduction in CA1 neurons, accompanied by changes in neuronal nitric oxide synthase (nNOS) and β secretase levels.[17] In another study, the intraperitoneal co-administration of aluminum chloride and D-galactose for one week in male Wistar rats induced depressive and anxiety-like behaviors, along with cognitive impairments.[18]

Moreover, chronic aluminum exposure disrupts antioxidant defense mechanisms, increases lipid peroxidation, and depletes endogenous antioxidants. In rats, daily oral administration of aluminum for 30 days altered synaptic neurotransmission by affecting key enzymatic activities, including acetylcholinesterase and monoamine oxidase (MAO).These alterations led to the depletion of neurotransmitters such as dopamine and serotonin, while excitatory neurotransmitters like glutamate and norepinephrine accumulated abnormally.[19]

In addition to its neurotoxic effects, aluminum exposure influences neurotrophic factor expression. Studies in rats have demonstrated reduced gene and protein levels of brain-derived neurotrophic factor (BDNF), nerve growth factor (NGF), glial cell line-derived neurotrophic factor (GDNF), and neurotrophin-3 (NT3) following aluminum exposure.[20] Cardiometabolic disturbances have also been reported in animals exposed to aluminum chloride, including alterations in glycemic and lipid profiles, reduced insulin levels, and abnormal electrocardiogram (ECG) patterns.[21] Age-related differences in aluminum toxicity have been explored, revealing that aged rats exhibit greater vulnerability. In older animals, aluminum chloride exposure led to an increase in hyperchromic neurons in the hippocampus, microglial activation, elevated pro-inflammatory cytokines, and markers of cellular senescence. Simultaneously, there was a reduction in anti-inflammatory cytokines and HIF-1α levels. Peripheral blood analysis indicated alterations in immune cell composition in aged rats exposed to aluminum chloride.[22] Furthermore, these animal models have provided insights into gut microbiota dysregulation in AD, linking aluminum exposure to changes in the gut–brain axis.[23,24] Biochemical alterations have also been reported in multiple organ systems. In mice, oral administration of aluminum chloride resulted in increased β-secretase and tau protein expression in the brain. In the liver, elevated alanine aminotransferase (ALT) and aspartate aminotransferase (AST) activities indicated hepatic damage, while upregulation of mitogen-activated protein kinase (MAPK) and c-Jun N-terminal kinase (JNK) suggested cellular stress and inflammation. In the kidneys, elevated levels of kidney injury molecule-1 (KIM-1) were observed, indicating renal toxicity.[25]

The summarized features of aluminum exposure are presented in Table 1.1.

1.3.1.2 Amyloid-beta

The intracerebral injection (mostly intracerebroventricular or intrahippocampal) of Aβ1–42 is one of the most widely used animal models of AD. Aβ is central to the amyloid cascade hypothesis, and its fibrillar form constitutes the primary component of amyloid plaques found in the brains of AD patients.

Aβ peptides are generated from the APP, which plays a central role in AD pathogenesis. This integral membrane protein is expressed in many tissues, particularly sensory synapses, and consists of a single membrane-spanning domain, a large extracellular glycosylated N-terminus, and a shorter cytoplasmic C-terminus. APP is implicated in synaptic formation, repair, anterograde neuronal transport, and iron export.

It is primarily cleaved by β-secretases and γ-secretases to produce Aβ, a 37–49 amino acid residue peptide. The amyloidogenic processing of APP involves consecutive cleavages by β-secretase and γ-secretase at the N and C termini of Aβ, respectively. The 99-amino-acid C-terminal fragment of APP, called C99, generated by β-secretase cleavage, undergoes internalization and further processing by γ-secretase at multiple sites. This results in cleavage fragments of various lengths (43, 45, 46, 48, 49, and 51 amino acids), which produce main Aβ forms: Aβ40 (containing 40 amino acids) and Aβ42 (containing 42 amino acids), within endocytic compartments.

TABLE 1.1
Key Characteristics of Aluminum Exposure in Rodents

Aspect	Details
Injection methods	Intracerebral, intracisternal, subcutaneous, intraperitoneal, and oral exposure
Effects	• Altered neurotransmitter levels • Cardiometabolic disturbances • Disruption of neurotrophic factors • Depressive and anxiety-like behaviors • Endoplasmic reticulum stress • Formation of neurofibrillary tangles and Aβ deposition • Gut microbiota dysregulation • Impaired learning, memory, and synaptic changes • Mitochondrial damage • Multi-organ toxicity • Neuroinflammation • Oxidative stress

Aβ monomers can aggregate into different structures, including oligomers, protofibrils, and amyloid fibrils. Amyloid fibrils are larger and insoluble, potentially forming amyloid plaques, while amyloid oligomers (AβO) are soluble and can spread throughout the brain.[26] Increasing evidence suggests that the AβO plays a critical role in initiating toxicity in AD pathogenesis. AβO has been demonstrated to be more neurotoxic compared to the Aβ fibrils.

Since 1990, Yankner et al. have proposed that Aβ fragments (amino acids 25–35) are neurotoxic to cultured rat cortical and/or hippocampal neurons. Consequently, primary culture systems have been used as an in vitro model to investigate the effects of Aβ.[27]

In vivo experiments have shown that isolated plaque cores from post-mortem AD brains[28], as well as synthetic Aβ fragments, can induce neuronal and neuritic loss, along with the expression of AD antigens, when injected into the rat brain. According to the evidence, AβO diffuse largely in the brain within 1 hour and up to seven days after injection.[29]

Aβ25–35 is identified as the core neurotoxic fragment of the full-length Aβ1–42 peptide. It represents the shortest Aβ sequence capable of forming β-sheet-aggregated structures with biological activity comparable to full-length Aβ. Truncated Aβ25–35 fragments are detected in the brains of AD patients but not in age-matched controls. They have been demonstrated to induce progressive pathology in rodent brains. Utilizing the Aβ25–35 fragment facilitates the study of Aβ toxicity due to its ability to induce toxicity similar to full-length Aβ fragments, while its shorter length allows for improved diffusion and solubility.

Inflammation is one of the features of AD that is induced following Aβ injection. Three days after the injection of Aβ1–42, immediate neuroinflammatory and vascular reactions were observed in the CA1 subdivision of the hippocampus, preceding noticeable neuronal degeneration. By the seventh day, a significant decrease in pyramidal cells within the CA1 area of the hippocampus was noted. By the 30th day, mice injected with Aβ1–42 exhibited a substantial increase in the expression of p-tau, GFAP, Iba-1, CD31, and alpha-smooth muscle actin (α-SMA) compared to the control group.[30]

Increased expression of inflammatory markers such as tumor necrosis factor-alpha (TNF-α) and IL-6 has been reported 14 days after Aβ administration.[31] Seven days after the injection of Aβ1–42 into the rat hippocampus, there was a notable increase in the levels of inflammatory cytokines, including interferon-gamma (IFNγ) and IL-1β[32]

It seems that administration of Aβ induces age-dependent neuroinflammation and cholinergic loss. Injection of Aβ25–35 into the lateral ventricles resulted in an age-related increase in cholinergic loss and microglial activation in the basal forebrain, accompanied by neuronal loss in the hippocampal CA3 subfield. Morris water maze performance demonstrated deficits in long-term reference memory in older animals exposed to Aβ, whereas younger animals did not exhibit this impairment.[33]

Inflammation has also been observed in the white matter tracts of rats that received icv AβO.[34] After intracerebroventricular (icv) injection of Aβ1–42, there is an increased expression of inflammatory markers detected in CSF. Additionally, the injection alters the morphology and integrity of choroid plexus epithelial (CPE) cells, leading to an increase in the blood–CSF barrier (BCSFB) leakage.[35] It has been shown that the action of AβO on memory was abolished by a selective Toll-like receptor 4 (TLR4) receptor antagonist. In TLR4 knockout mice, it did not have any effect on either memory or glial activation.[36] The findings suggest that higher levels of inflammatory markers in the hippocampus following Aβ injection could also send signals to the rest of the body, leading to changes in the overall immune responses.[37]

Another proposed mechanism of action for intracerebrally injected Aβ involves cysteinyl leukotrienes, which are inflammatory lipid mediators originating from the 5-lipoxygenase pathway of arachidonic acid metabolism. Their potential role in the pathogenesis of AD has been evidenced.[38]

The impairment in the synthesis of proteins related to the cholinergic system, such as choline acetyltransferase (ChAT) and vesicular acetylcholine transporter (VaChT), is also highlighted as one of the crucial mechanisms in amyloid-induced neuronal damage.[39]

In a study examining molecular activities associated with apoptosis, autophagy, and mitophagy following Aβ1–42 injection over ten days, researchers noted various trends. The cleavage of caspase-3 began to increase slightly three days after the Aβ injection. Autophagy markers peaked and then returned to baseline levels as apoptotic markers started to rise. Mitophagy markers exhibited a similar pattern to autophagy markers, indicating concurrent removal of damaged mitochondria. Researchers observed a shift toward mitochondrial fission in mitochondrial dynamic processes, suggesting impaired mitochondrial generation in response to Aβ. Additionally, the activity of specific enzymes involved in the antioxidant defense system, electron transport chain, and tricarboxylic acid cycle (TCA) decreased in response to Aβ, alongside the suppression of mitochondrial biogenesis.[40]

icv injection of oligomeric Aβ25–35 induces activation (phosphorylation) of P38MAPKs, elevates the level of cleaved caspase-3, and decreases the number of neurons in the rat hippocampal CA1 and CA3 subfields.[41]

Studies have demonstrated that in addition to the accumulation of Aβ peptides, soluble forms of the Aβ peptide also play a role in cognitive impairment and interfere with synaptic mechanisms associated with memory. When soluble Aβ peptide is injected icv, it disrupts the consolidation and retrieval of long-term memory. This effect of soluble Aβ might suggest the involvement of the glutamatergic system.[42]

Impairments in hippocampal-dependent habituation and recognition memory have been reported.[43] Studies have demonstrated that the administration of Aβ1–42 resulted in cognitive impairments observed three days post-injection.[44] Disrupted functional connectivity in the prefrontal cortex (PFC) during the assessment of working memory has also been observed after Aβ injection.[45]

The coordination between spikes and local field potentials (LFPs) recorded from the prefrontal cortex, which contributes to working memory–related cognitive function, is weakened after 14 days following the administration of Aβ1–42 potentially contributing to the cognitive deficits observed in AD.[46]

The repeated intra-CA1 injection of Aβ25–35 over four days was also found to induce AD-like behavioral impairments in the rats. The authors claimed that these repeated injections produce a more reliable model for AD compared to a single-dose injection.[47]

Inhibition of long-term potentiation (LTP) in the dentate gyrus region has been reported following acute treatment with Aβ1–42.[48] icv administration of Aβ enhances the excitability of the CA3–CA1 synapse and disrupts LTP and hippocampal oscillatory patterns.[49]

Aβ1–42 leads to reduced hippocampal expression of BDNF, a neurotrophic factor crucial for learning and memory. This modulation is influenced by the activity of histone deacetylase 2 (HDAC2).[50]

Moreover, an aberrant synchronization of neural activity within the hippocampal network has been detected, signifying disrupted inter-neuronal communication, potentially contributing to memory deficiencies.

Asymmetric loss of various synaptic markers has been reported after Aβ treatment, with glutamatergic synapses showing a predominant susceptibility.[51] Also, Aβ1–42 reduces the expression of glutamatergic receptors and transporters in specific brain regions and layers.[52]

Intrahippocampal administration of Aβ1–40 disrupts silent synapses, potentially due to dysfunctions in the regulation of actin cytoskeleton dynamics and postsynaptic scaffolding proteins.[53]

SV2B, a synaptic protein found extensively throughout the brain, plays a role in the intricate machinery of synaptic vesicle regulation, including endocytosis and exocytosis. It is suggested to be one of the crucial regulators of amyloid toxicity at the synaptic level.[54] The administration of Aβ also affects the neurotransmitter system. In the studied region, an increase in norepinephrine concentration was detected 2 hours after an Aβ icv injection. It is suggested that this change is mediated by NO and IL-1.[55]

In AD, there is early dysregulation of the hypothalamic–pituitary–adrenal (HPA) axis, leading to abnormal levels of circulating glucocorticoids and impaired signaling of glucocorticoid receptors.

icv injection of Aβ leads to an elevation in glucocorticoid levels in plasma and results in synaptic deficits, apoptosis, and neuroinflammation in PFC, a cerebral area abundant in glucocorticoid receptors. The PFC is recognized to be impacted at an early stage of AD and is crucial in regulating the HPA axis as well as cognitive processes.[56]

Additionally, Aβ can induce depressive-like behavior in rats, along with reduced cortical 5-HT and neurotrophin levels. The antidepressant treatment was able to reverse this phenotype. Additionally, in the same model, a dysfunction in the HPA axis linked to increased cortical noradrenaline levels was observed.[57] Metabolic disturbances and weight loss have also been reported as a consequence of Aβ injection.[58]

Aβ can also induce peripheral metabolic dysregulation by targeting the hypothalamus. For example, peripheral glucose intolerance has been observed following the icv infusion of AβO.[59] Also, Aβ infusion in the lateral ventricle alters rats' ability to acclimate to heat and modifies their behavioral thermoregulation.[60]

The low-density lipoprotein receptor (LDLr), has been identified as a crucial receptor in the central nervous system (CNS) that regulates the deposition of Aβ in mouse models of β-amyloidosis. In this context, studies have demonstrated that mice lacking LDLr, irrespective of Aβ treatment, displayed increased permeability of the BBB and memory deficits. Additionally, LDLr knockout mice treated with Aβ1–40 peptide exhibited elevated levels of acetylcholinesterase activity, astrogliosis, oxidative imbalance, and increased cell permeability in the hippocampus compared to Aβ-treated wild-type mice. Therefore, the lack of LDLr heightens vulnerability to Aβ-induced neurotoxicity.[61]

A research study examined the effects of icv administration of Aβ 1–42 on the oxidative stress parameters and antioxidant defenses in the prefrontal cortex of rats over 24 hours. The findings indicated a disturbance in the typical temporal patterns of the cellular redox status within the prefrontal cortex of rats.[62]

It is suggested that extracellular Zn^{2+} may play a crucial role in transiently inducing Aβ-induced cognition deficits. Studies have reported that upon injection of Aβ into the DG, intracellular Zn^{2+} levels increased specifically in the injected area. This finding indicates that Aβ induces the influx of Zn^{2+} into cells within the injected region.[63,64]

An increased formation of NO and lipid peroxidation was also reported two weeks after the injection of Aβ 25–35 in the brain, as well as in the kidneys and liver of the animals.[65]

It has been observed that the activation of ferroptosis and necroptosis pathways occurs in an animal model induced by Aβ.[66]

It has been shown that injected Aβ peptides can be transmitted through neural connections throughout the brain. When Aβ is infused into the olfactory bulb, it swiftly and easily transfers from the injection site to interconnected brain regions through neural connections, leading to neuronal apoptosis in the affected areas.[67]

Research has shown that Zn^{2+} plays a critical role in promoting the absorption of Aβ1–42 into dentate granule cells within the hippocampus of rats.[68] A study reported that intracellular disruption of Zn^{2+} homeostasis triggered by Aβ1–42 predominantly leads to neuronal degeneration within the dentate gyrus, consequently leading to impairment in cognitive functions dependent on the hippocampus.[69] It is important to note that rat Aβ1–42 appears to exhibit reduced binding affinity toward extracellular Zn^{2+} compared with human Aβ1–42. Consequently, rat Aβ1–42 is unable to capture Zn^{2+} in the extracellular space, resulting in no significant impact on cognitive function in rats.[70]

icv injection of Aβ1–42 leads to a simultaneous increase in CSF and plasma levels of Aβ in a dose-dependent manner. The authors of this study posit a correlation between Aβ concentrations in both CSF and plasma during the acute phase of this model.[71] An interesting study showed that the Aβ-inducing activity of AD brain tissue remains preserved, albeit with reduced effectiveness, following formaldehyde fixation. The durability of Aβ seeds, which resist inactivation and structural changes induced by formaldehyde, underscores their resilience and may facilitate their persistence and widespread distribution throughout the body.[72]

Furthermore, it has been demonstrated that icv injection of Aβ1–42 results in the initial activation of neurogenic niches, but neural stem cells fail to complete the maturation cascade, resulting in neural stem cell immaturity.[73]

Additionally, the induction of ER stress has been documented as a plausible molecular process involved in the development of AD-like neuropathology.[74] After icv injection of Aβ1–42, there have been reports of decreased expression of peroxisomal proteins such as PEX14, CAT, and PMP70.[75]

Also, the administration of Aβ can elicit an imbalance in the gut microbiota, change the structure of the intestines, and accelerate the processes of amyloidogenesis.[76]

During SWS, post-learning sharp wave ripples (SWRs) in the hippocampus are believed to have a crucial role in memory formation. It has been reported that mice receiving Aβ experience a disruption of SWR dynamics.[77]

After injecting Aβ1–42 aggregates, researchers have observed significant changes in the daily oscillation of protein carbonyls, disturbances in the rhythms of clock genes, and varying effects on the expression and activity of antioxidant enzymes. These changes were monitored in samples collected from the hippocampus at regular intervals over 24 hours.[78]

In summary, the intracerebral injection of Aβ1–42, including its various truncated fragments such as Aβ25–35, has proven to be a robust and widely utilized animal model for AD. Despite substantial progress, further exploration of the molecular mechanisms and potential therapeutic targets remains essential. The Aβ-based AD models will undoubtedly continue to play a central role in identifying novel strategies for intervention, with the hope of alleviating the burdens imposed by this devastating neurodegenerative disorder.

The summarized effects of Aβ injection in rodents are presented in Table 1.2.

1.3.1.3 Scopolamine

Scopolamine, a muscarinic cholinergic receptor antagonist, is widely used to induce cognitive impairment in experimental animal models. By blocking muscarinic receptors in the cerebral cortex, scopolamine disrupts cholinergic signaling, leading to excessive acetylcholine release.

TABLE 1.2
Key Characteristics of Amyloid-Beta Exposure in Rodents

Aspect	Details
Injection methods	Intracerebroventricular or intrahippocampal
Effects	• Amyloid plaque formation • Anxiety-like behavior, depressive symptoms, and altered activity levels • Cholinergic deficits • Circadian rhythm disruption • Cognitive impairment • Gut microbiota alterations • Mitochondrial dysfunction • Neuronal death • Neuroinflammation • Oxidative stress • Synaptic dysfunction • Tau protein hyperphosphorylation

This imbalance contributes to neurotoxicity in the hippocampus, resulting in dose-dependent deficits in learning and memory, particularly in mice.

The central effects of scopolamine are primarily attributed to the blockade of M1 and, to some extent, M5 muscarinic receptors, given their distinct distribution in the brain. However, scopolamine-induced memory impairment may involve N-methyl-D-aspartate (NMDA) receptor mechanisms, particularly in the dorsal hippocampus and septum .[79]

The scopolamine model of aging and dementia serves two key purposes: first, to illustrate how central cholinergic dysfunction can induce cognitive impairments resembling those seen in aging and AD; and second, to provide a platform for evaluating potential anti-dementia drugs based on their ability to counteract the cognitive deficits caused by cholinergic blockade.

Unlike traditional cholinergic models of cognitive impairment, the scopolamine model is distinctive in its reversibility by drugs that enhance cognition through non-cholinergic mechanisms, highlighting its versatility in investigating various pathways involved in cognitive function.

For modeling AD, scopolamine can be administered either intracerebrally or systemically.

Subcutaneous injection of scopolamine has been shown to impair various cognitive domains in animals, such as deficits in visual recognition memory, visuospatial praxis, visuospatial recall, and psychomotor speed.[80] Additionally, intraperitoneal injection of scopolamine has been found to disrupt both contextual and cued fear conditioning in both young and aged mice, with effects observed as early as 48 hours post-administration.[81]

Hippocampal infusions of scopolamine inhibit LTP[82] and affect spatial memory encoding.[83] Given the role of muscarinic receptors in the central regulation of executive function, evidence suggests that scopolamine impairs both affective (reversal learning) and attentional set-shifting components in rats.[84]

Scopolamine disrupts synaptic plasticity through changes in HDAC2 and DNA methyltransferase-1 (DNMT1), leading to impaired synaptic plasticity gene expression and memory consolidation.[85]

Scopolamine was found to induce impairments in spatial learning and memory through the activation of glycogen synthase kinase-3 beta (GSK-3β), leading to disturbances in spine formation/maturation and dendrite arborization, and the expression of key synaptic proteins such as CREB, Homer1, and α-amino-3-hydroxy-5-methyl-4-isoxazole propionic acid receptor (AMPA-R).[86]

Also, systemic administration of scopolamine has been shown to impair passive avoidance memory, an effect correlated with decreased levels of matrix metalloproteinases (MMP-2 and MMP-9) and increased activation of MAPKs in the hippocampus. Additionally, this model is associated with diminished phosphorylation of Akt and GSK-3β, contributing to disruptions in memory retrieval.[87]

Modulation of specific receptor systems has been shown to counteract scopolamine-induced cognitive deficits. Activation of the angiotensin AT_4 receptor effectively improves spatial working memory impairment caused by intrahippocampal scopolamine, suggesting its crucial role in cognitive processes.[88] Similarly, sigma (σ1) receptor-selective agonists exert potent neuroprotective effects, significantly ameliorating scopolamine-induced memory deficits.[89] Additionally, the blockade of 5-HT receptors has demonstrated protective effects against scopolamine-induced memory impairment, underscoring the involvement of serotonergic signaling in mitigating cholinergic dysfunction.[90,91]

Furthermore, gene expression analyses in scopolamine-treated animals have identified Homerl and GABA(B) receptors as key contributors to cognitive impairment, alongside calcium/calmodulin-dependent protein kinase 2 (CaMKII), further implicating GABAergic and glutamatergic dysregulation.[92]

Scopolamine administration also disrupts metabolic function, with notable alterations in L-arginine metabolism within the hippocampus and prefrontal cortex. Specifically, the intraperitoneal administration of scopolamine leads to significant changes in the activity of NOS and arginase enzymes. These changes are accompanied by fluctuations in the levels of arginine and its metabolites, particularly within the dentate gyrus sub-region, highlighting the metabolic disturbances associated with scopolamine-induced cognitive impairment.[93]

Scopolamine treatment has been shown to impair mitochondrial function in rats, manifested by a loss of mitochondrial membrane potential ($\Delta\Psi$m) and the induction of cytochrome c release in brain tissue. This mitochondrial disruption contributes to cognitive impairment and highlights a crucial aspect of the pathophysiology in scopolamine-induced models of memory impairment.[94,95]

In addition to mitochondrial dysfunction, oxidative stress, and neuroinflammation are significant contributors to cognitive decline in this model. In scopolamine-treated rats, a marked reduction in the activities of key antioxidant enzymes, including superoxide dismutase (SOD), glutathione, peroxidase, and catalase, was observed. This decrease in antioxidant enzyme activity coincides with a reduction in the overall content of reduced glutathione in the hippocampus. The imbalance in oxidative homeostasis is further indicated by a significant increase in the production of malondialdehyde, a marker of lipid peroxidation.[96] Moreover, scopolamine administration induces neuroinflammation, as evidenced by elevated levels of pro-inflammatory cytokines, such as IL-1β, IL-6, IL-17, and interferon-gamma (IFN-γ), further exacerbating cognitive deficits.[97]

Additionally, alterations in key markers of neuroplasticity were observed. Two and four weeks following systemic scopolamine administration, a decline in myelin basic protein expression, as well as reductions in BDNF and tropomyosin receptor kinase B (TrkB) levels in the dentate gyrus of the hippocampus, suggests impaired neuroplasticity and synaptic function in this model.[98]

In line with other molecular changes, chronic systemic administration of scopolamine over extended periods (6 and 12 weeks) in male Wistar rats has been shown to result in increased Aβ deposition in the hippocampus. This increase could be attributed to elevated expression and activity of β-site APP cleaving enzyme 1 (BACE1).[99]

In parallel, in vivo electrophysiological assessments have demonstrated that scopolamine significantly reduces the power of elicited theta oscillations in the hippocampus of anesthetized animals. This reduction in theta rhythm activity suggests compromised hippocampal function, further corroborating the cognitive deficits induced by scopolamine.[100]

Furthermore, scopolamine triggers the activation of apoptotic signaling pathways, contributing to neuronal cell death. The injection of scopolamine leads to an increase in TUNEL-positive cells, a higher Bax/Bcl-2 ratio, as well as enhanced caspase-3 activity and cleavage of poly(ADP-ribose) polymerase (PARP), all of which are indicative of apoptotic cell death in brain tissues.[95]

TABLE 1.3
Key Characteristics of Scopolamine Exposure in Rodents

Aspect	Details
Main mechanism	Muscarinic cholinergic receptor antagonist
Injection methods	Intraperitoneal, subcutaneous, or intracerebral injections
Effects	• Amyloid plaque formation • Cognitive impairments • Spatial learning and memory deficits • Metabolic dysfunction • Mitochondrial dysfunction • Neuroinflammation • Neuronal death • Oxidative stress • Synaptic and plasticity changes

The scopolamine-induced memory impairment model is widely utilized due to the absence of mandatory surgical procedures. Given that cholinergic neurons and muscarinic receptors are distributed throughout the CNS, scopolamine disrupts memory tasks through various pharmacological mechanisms, making it an invaluable tool for studying cognitive impairment in the context of AD and related disorders.[101]

The mechanisms of action, injection techniques, and resulting effects of scopolamine administration in rodents are summarized in Table 1.3.

1.3.1.4 Streptozotocin

Streptozotocin (STZ) is a glucosamine-nitrosourea compound originating from soil bacteria called Streptomyces achromogenes, first identified in the late 1950s.

Originally approved as an anti-neoplastic agent, subsequent research discovered its toxicity to β-cells, the pancreatic cells responsible for insulin secretion. Since then, peripheral injection of STZ (around 40–60 mg/kg) has been commonly used to model diabetes in experimental studies.

The selective toxicity of STZ toward β-cells may be due to its glucose component, enabling entry into cells via glucose transporter type 2 (GLUT-2).[102]

GLUT-2 is widely distributed throughout the rat brain, particularly in limbic areas and related nuclei, with the highest concentrations found in the ventral and medial regions near the midline. It plays a crucial role in glucose sensing and the regulation of neurotransmitter release. The pathological changes induced by central injection of STZ occur independently of β cell impairments.

STZ shares structural similarities with glucose and N-acetylglucosamine (GlcNAc), and pancreatic β-cells are selectively sensitive to STZ due to their higher levels of O-GlcNAc-selective N-acetyl-β-D-glucosaminidase (OGlcNAcase), an enzyme responsible for attaching O-GlcNAc to proteins, compared to other cell types. STZ has been shown to inhibit the enzyme OGlcNAcase, leading to cytotoxicity specifically in β-cells.

In the brain, another toxicity mechanism is proposed. STZ is a methylating or DNA-alkylating agent that forms carbonium ions (CH^{3+}) inducing DNA damage through DNA methylation. It also instigates DNA fragmentation due to the nitrosourea moiety, leading to neuronal death.[103]

Professor Sigfried Hoyer is recognized as the pioneer of the icv-STZ model. His early contributions to this field were summarized in a paper published in 1994.[104] The reasoning developed by Hoyer begins with the observation that in sporadic AD, the brain shows decreased consumption of oxygen and glucose. This led to the suggestion that the primary biochemical perturbation in earlystage sporadic AD revolves around the regulation of cerebral glucose metabolism. This deregulation is believed to result from a failure in cerebral insulin receptor signaling.[104] Another observation considered at this point was that icv insulin administration stimulated hexokinase and phosphofructokinase by approximately 20% in adult male Wistar rats. This finding highlighted the significance of insulin signaling in regulating brain glucose metabolism, particularly in controlling glycolytic flux. Therefore, it was hypothesized that any disruption in insulin signal transduction would significantly impact brain glucose metabolism.[105] As a next step, STZ was injected intracranially to induce damage to glucose and energy metabolism in the brain. In the early stages of sporadic AD, reductions in glucose metabolism in the cortex and impaired cerebral blood flow are commonly observed. Following icv-STZ administration, severe abnormalities in brain glucose/energy metabolism, along with reduced cerebral blood flow, have been documented. Specifically, reduced glucose utilization has been observed in 17 out of 35 brain areas, primarily affecting the frontal, parietal, auditory, sensory–motor, and entorhinal cortex, as well as all hippocampal subfields.[106]

Compelling evidence indicates that single or multiple (mostly two) low-dose injections of STZ (1–3 mg/kg), either uni- or bilaterally into the lateral cerebral ventricles, induce reduced cerebral blood glucose levels, neurochemical changes, and behavioral alterations resembling those seen in AD.[107]

It should be noted that in addition to the lateral ventricles, bilateral injection of STZ into the hippocampus of rats also results in cognitive impairment and other cellular and molecular hallmarks of AD.[108]

It is believed that the development of Alzheimer-like symptoms in rats after one or two icv-STZ injections is a dynamic process. Most studies have observed the effects of icv-STZ on brain function and behavior over a few weeks. However, some studies have investigated longer follow-up periods in STZ-icv–treated rats.

Some believe that the characterization of the STZ-icv model has been limited to a duration of up to three months following STZ-icv administration, which is insufficient to replicate the human disease process, typically unfolding over several decades. However, evidence shows signs of sporadic AD even after nine months of STZ injection. Cognitive deficits exhibit a distinct time pattern, with an acute decline observed after one month and a slow but progressive chronic decline evident from three months following STZ-icv treatment. It is plausible to speculate that the intensive acute response observed after one month may reflect the toxic effects of STZ, which could be partially compensated for until three months after the STZ-icv treatment. In addition, there is evidence of diffuse Aβ1–42 plaque-like formations, accumulation of hyperphosphorylated tau protein, and ultrastructural changes, all of which develop in a time- and dose-dependent manner over the course of nine months.[109]

Also, CAA persists and progresses in this animal model, evident even nine months following STZ-icv treatment, as an extension of the amyloid pathology initially observed three months post-treatment.[110]

It appears that cognitive and neurochemical changes induced by STZ-icv injections follow a time-dependent pattern consisting of three phases. Within the first month, an acute response develops, followed by a tendency to return to normal values in one to three months. Finally, at six to nine months, a decompensation phase occurs with a slow and progressive aggravation. This temporal progression of pathology initiated by STZ injection into the lateral ventricles, whether single or double, mirrors real-life situations, unlike transgenic AD models.[111]

In a study comparing the short-term and long-term effects of STZ on adult hippocampal neurogenesis, the results indicated that while STZ-icv treatment did not significantly influence cell proliferation over a short-term period (one month), it significantly decreased the generation of immature and mature neurons in the long term (three months). This reduction observed after three months was specific to the septal hippocampus, which is considered important for spatial learning.[112]

STZ administration to the rodent brain induces an insulin-resistant brain state, gradual cholinergic deficits, oxidative stress, neuroinflammation, and other biochemical changes that are considered early pathological features observed in AD. Several molecular studies conducted on rodents administered STZ revealed its potency to alter both mRNA transcripts and proteins of genes involved in insulin signaling pathways, as well as genes related to AD such as tau, APP, and BACE.[113]

The administration of STZ resulted in impairment of both short-term and long-term memory, as assessed after 14 and 21 days, respectively. At the 14-day mark, the STZ group displayed significant alterations in the majority of the evaluated AD markers compared to the control group. These alterations included increased expression of Aβ protein and neurofilaments, elevated phosphorylation of Tau protein, and decreased levels of synapsin compared to the control group. Notably, these neurochemical changes were transient and did not persist beyond 21 days following the STZ injections, except for synapsin.[114]

Disruption of working memory was observed as early as 3 hours after STZ injections, a finding that could be correlated with observations indicating that patients in the earliest stages of AD often display mild impairments in short-term memory.[115] STZ induces both acute and prolonged neuroinflammation, accompanied by a significant decrease in neural stem cell proliferation in the hippocampal dentate gyrus and subventricular zone. Furthermore, it diminishes the survival, differentiation, and maturation of newborn neurons, ultimately resulting in deficits in short-term and long-term spatial memory.[116]

The STZ model not only affects cognitive function but also has an impact on non-cognitive behaviors, specifically anxiety-like behavior.[117]

Another significant characteristic of the rat STZ model, mirroring observations in the brains of AD patients, involves indications of oxidative stress in brain tissue and abnormalities in brain mitochondria, potentially leading to apoptotic cell death mediated by caspases.

In a study conducted five weeks after a single icv injection of STZ (3 mg/kg), brain mitochondria exhibited deficits in function. These impairments included a decrease in mitochondrial transmembrane potential, ATP content, and repolarization level, along with an increase in the lag phase of repolarization. Furthermore, mitochondria showed reduced activities of pyruvate and α-ketoglutarate dehydrogenases and cytochrome c oxidase. Additionally, there was an elevation in hydrogen peroxide and lipid peroxidation levels, coupled with a reduction in glutathione content observed in mitochondria from STZ-icv rats.[118]

Some researchers hypothesize that STZ-icv infusion leads to protein aggregation and dementia, possibly due to increased mitochondrial fragmentation and functional abnormalities. In a 60-day follow-up study, dynamic changes in mitochondrial function have been reported. Following STZ-icv infusion, mitochondrial electron transport chain complex-I activity significantly decreased in the hippocampi and cortices within 14 days, persisting for up to 60 days. Mitochondrial state-3 respiration remained unchanged initially but decreased significantly from 21 days post-STZ administration. Dynamin-related protein 1 (DRP-1) expression markedly increased in the hippocampi and cortices by 21 days post-infusion, persisting only in the hippocampi for up to 60 days.[119]

Significant brain atrophy and ventricular enlargement have been reported in STZ-treated animals. In addition, weight loss is frequently observed following STZ treatment, suggesting a possible relationship between body-weight loss and the decrease in mitochondrial complex I–III activity in the brain after STZ treatment.[120]

Using magnetic resonance imaging (MRI) and positron emission tomography (PET) scan in a longitudinal study, altered functional connectivity, intra-axonal damage, and demyelination in AD-related brain regions in icv-STZ rats have been found.[121]

Also, STZ induces specific damage to myelinated tracts in the fornix by generating oxidative stress, consequently disrupting connections between the septum and hippocampus.[122]

It has been shown that STZ is associated with broad and sustained abnormalities in the white matter of the temporal lobe.[123] STZ administration also impacted cellular communication, evidenced by an increase in the expression of connexin 43.[124]

Additionally, STZ treatment induces endoplasmic reticulum (ER) stress. Elevated calcium levels were observed in both the cortical and hippocampal regions of rat brains after STZ treatment.

The increased calcium levels may be related to ER stress, as indicated by significantly heightened expression of ER stress markers such as glucose-regulated protein 78 (GRP78), growth arrest and dna damage (GADD), and caspase-12 in the brains of STZ-treated rats.[124]

It appears that the STZ model inadequately mirrors central cholinergic dysfunction, which is a crucial characteristic of dementia.[125]

In a study examining the cholinergic system after 30 and 120 days, it was found that cholinergic system disruption was observed only in the hippocampus and only after 30 days. The authors concluded that this suggests the STZ-induced animal model may not be the most suitable for studying the cholinergic system in AD, or that it may only yield informative results for a limited duration.[126]

Most observations align with Hoyer's findings that metabolic disturbances following STZ-icv adminstration spread across extensive regions of the brain, including areas distant from the injection site, and progressively accumulate over time. For instance, three weeks after STZ injection, a significant enlargement of the Golgi apparatus occurred in the frontoparietal cortex due to the expansion of the trans-Golgi segment of the cellular protein secretory pathway.[127] After six weeks of administering STZ, breakdown in the phospholipids of the membrane and damage to the membrane structure were reported. Accordingly, the levels of palmitic acid and stearic acid were found to be higher in the temporal cortex of the brain. Additionally, the concentration of palmitic acid was also increased in the entorhinal cortex.[128] Four weeks after STZ treatment, rats exposed to STZ showed significantly lower catalase (CAT) activity in the hippocampus, brain stem, and cerebellum compared to the control group.[129]

According to recent evidence, the gut is implicated in the development and progression of AD by triggering systemic inflammation and oxidative stress. Interestingly, a dysfunctional gut–brain axis as well as pathophysiological changes in the gut were reported in this model, showing that it can be used to study CNS-related gastrointestinal consequences. Gastrointestinal barrier dysfunction, altered mucus composition, redox dyshomeostasis, and altered absorption are characteristic features of the STZ-icv model.[130–132]

Sleep disturbances are also common in individuals with AD and frequently occur before the onset and advancement of dementia. Rats treated with STZ-icv showed gradual reductions in SWS during the light phase and across the light/dark cycle starting from 7 days after treatment. Furthermore, increased wakefulness and decreased REM and non-REM (NREM) sleep were noted from 14 days following STZ-icv treatment. Using immunohistochemical staining, it was revealed that the parafacial zone (PZ) lost a significant number of GABAergic neurons that were c-Fos-positive by day seven after STZ. So that according to these findings, the STZ-icv rat model is useful for evaluating sleep disturbances associated with AD.[133]

Respiratory dysfunction is also reported following STZ treatment. Significant atrophy was observed in the caudal region of the nucleus tractus solitarii in rats treated with higher doses of STZ. This region is responsible for integrating chemoafferent information.[134]

STZ-icv adminstration has been shown to sensitize both central and/or peripheral receptors to pain. The effects of STZ on pain sensitivity vary depending on the dose and time of administration. It is suggested that the AD-like pathology induced by STZ may be partly mediated through pain-processing pathways.[135]

The neuropathological changes induced by STZ depend on the timing of the investigation and the specific brain regions studied. For example, the changes observed in the rat hippocampus were specific to that region of the brain, and a different pattern of changes was found in the frontal cortex, another brain region involved in cognitive functions.

Furthermore, previous findings have demonstrated that there is individual susceptibility to STZ treatment in rats.[109] A study examining sex differences found that male rats exhibit greater responsiveness than female subjects to STZ-icv administration, suggesting that females may not be optimal for the STZ model of AD.[136]

A study on strain diversity found that Long-Evans rats, known for their superior cognitive abilities compared to other strains, are less sensitive to STZ treatment than Wistar rats. Higher doses may be required to induce pathological changes in Long-Evans rats.[137] So, variations in the age,

TABLE 1.4
Key Characteristics of Streptozotocin Exposure in Rodents

Aspect	Details
Origin	A glucosamine-nitrosourea compound from Streptomyces achromogenes, identified in the late 1950s.
Injection methods	Intracerebroventricular or intrahippocampal
Effects	• Amyloid angiopathy • Amyloid plaque formation • Brain atrophy • Cholinergic deficits • Cognitive impairments • Disrupted neurotransmitter release • Dysfunctional gut–brain axis • Endoplasmic reticulum stress • Gut microbiota alterations • Gastrointestinal barrier dysfunction • Increased pain sensitivity • Mitochondrial dysfunction • Neuronal death • Neuroinflammation • Oxidative stress • Reduced cerebral blood flow • Reduced glucose metabolism in the brain • Respiratory dysfunction • Sleep disturbances • Tau protein hyperphosphorylation

strain, and experience level of the animals, along with differences in STZ dosage and food restriction regimen, may contribute to discrepancies in STZ effects. This underscores the necessity for more comprehensive studies using the STZ model to further confirm and specify its translational relevance to human AD.[138]

Streptozotocin origin, induction methods, and effects are summarized in Table 1.4.

1.3.2 Genetic Models

1.3.2.1 Amyloid Precursor Protein Mutation-Based Mice

1.3.2.1.1 Tg2576 Mice

The Tg2576 mouse model, which overexpresses human APP (hAPP) in APP695 with the "Swedish" mutation (K670N/M671L) (Table 1.5), is widely used to investigate the relationship between Aβ pathology and memory deficits.[139] In addition to neuronal expression, hAPP is detectable in astrocytes of young Tg2576 mice even before the onset of amyloid pathology. However, microglial cells and oligodendrocytes do not exhibit detectable hAPP expression in this model.[140]

- Pathophysiology and Disease Progression
  - **Aβ Accumulation and Plaque Formation**

    Aβ levels rise rapidly at approximately 6 months of age, followed by the formation of amyloid plaques between 9 and 12 months. By 12 months, diffuse plaques become evident, and between 12 and 23 months, there is a significant accumulation of diffuse plaques, NPs with amyloid cores, and biochemically detectable Aβ42 and Aβ40,

TABLE 1.5
Genetic Mutations in Alzheimer's Disease Mouse Models: APP, PS, and Tau Variants

Model	APP Mutations	PS Mutations	Tau Mutations
Tg2576	Swedish (KM670/671NL)	None	None
Tg2576 × PS1	Swedish (KM670/671NL)	PS1 M146L	None
APPswe/PSEN1ΔE9	Swedish (KM670/671NL)	PSEN1 ΔE9	None
5XFAD	Swedish (KM670/671NL), Florida (I716V), London (V717I)	PS1 M146L, PS1 L286V	None
P301S Tau	None	None	Tau P301S
rTg4510	None	None	Tau P301L
3×Tg-AD	Swedish (KM670/671NL)	PS1 M146V	Tau P301L

resembling levels found in AD patients.[141] The entorhinal cortex, one of the earliest brain regions affected in AD, exhibits increased excitability in Tg2576 mice as early as two to four months of age, preceding Aβ deposition and the onset of a few behavioral impairments.[142]

- **Neurogenesis Deficits**
  Neurogenesis deficits are also evident, with impaired subventricular zone neurogenesis occurring even at presymptomatic stages. Additionally, neurogenesis defects have been reported in the olfactory bulb, further highlighting early dysfunction in this model.[143]
- **BBB Dysfunction**
  Aβ homeostasis is regulated by ATP-binding cassette transporters ABCC1 and ABCB1, which mediate efflux, while the receptor for advanced glycation end (RAGE) products facilitates influx across the BBB. In Tg2576 mice, BBB integrity is disrupted, as evidenced by compromised tight junctions, reduced ABCC1 functionality, and decreased RAGE binding. These alterations may contribute to impaired Aβ clearance and subsequent amyloid accumulation.[144]
- **Neurotransmitter Alterations**
  Functional deficits in the dorsal hippocampus of Tg2576 mice lead to behavioral alterations characteristic of AD. These deficits correlate with the early degeneration of dopamine neurons in the ventral tegmental area, which can be partially reversed by L-DOPA treatment. Moreover, beginning at three months of age, pyramidal neurons in the ventral CA1 region exhibit altered excitability and firing patterns, coinciding with reduced levels of tyrosine hydroxylase, the rate-limiting enzyme in DA synthesis.[145] Additionally, granule cell dysfunction in this model is characterized by increased excitatory synaptic input, reduced inhibitory signaling, and muscarinic cholinergic dysregulation.[146]
- **Mitochondrial Dysfunction**
  Transcriptomic analyses of the hippocampus and retina in Tg2576 mice reveal significant downregulation of oxidative phosphorylation and mitochondrial pathways, highlighting widespread metabolic dysfunction.[147]

  The gastrointestinal tract is increasingly recognized as a source of oxidative stress contributing to AD pathogenesis. In presymptomatic Tg2576 mice, heightened gastrointestinal electrophilic tone is observed, characterized by increased lipid peroxidation and elevated manganese/iron SOD (Mn/Fe-SOD) activity, indicative of oxidative stress. As the disease progresses, these alterations partially normalize; however, total antioxidant capacity declines, suggesting a shift in redox balance.[148]

- **Gut Microbiota Dysbiosis**
  Tg2576 mice exhibit neuropathological constipation and gut microbiota dysbiosis, with a reduction in Bacteroidetes and an increase in Firmicutes and Proteobacteria at the phylum level.[149]
- **Network Hypersynchrony**
  In AD, there is an occurrence of network hypersynchrony that leads to epileptic activities. Studies have shown that epileptic events are more likely to happen during REM sleep in the Tg2576 mouse model. These epileptic events do not seem to worsen with age in this model. Additionally, it has been observed that epileptic events in REM sleep are closely synchronized with a specific part of the theta cycle. This synchronization with the hippocampal theta rhythm, where pyramidal cells have higher firing rates, could be a possible mechanism behind the timing of epileptic events. Furthermore, it has been revealed that there is an antiepileptic mechanism of noradrenergic transmission via α1-adrenoreceptors, which could explain the distribution of epileptic events over the sleep–wake cycle in this model.[150]
- **Peripheral Effects**
  Beyond central nervous system pathology, Tg2576 mice exhibit reduced cholinergic innervation of skeletal muscle, which may underlie AD-associated sarcopenia.[151] Furthermore, osteoporotic changes, including decreased bone formation and increased bone resorption, have been observed in this model, suggesting systemic manifestations of AD pathology.[152]

  In summary, the Tg2576 mouse model provides valuable insights into the pathophysiological mechanisms underlying AD. Its progressive Aβ accumulation, network hypersynchrony, neurotransmitter dysregulation, mitochondrial dysfunction, and peripheral manifestations make it a comprehensive model for investigating disease progression and potential therapeutic interventions.

1.3.2.2 Amyloid Precursor Protein and Presenilin Mutation-Based Mice

APP-Tg mice exhibit some classical pathological features of AD; however, they lack several other key components of the disease. Therefore, combining previously described AD models with other familial AD-associated mutations has been attempted to develop a more comprehensive model that can replicate the remaining pathological characteristics of AD. Presenilin-1 (PS1) and Presenilin-2 (PS2) genes encode subunits of γ-secretase, an enzyme crucial for processing APP and generating Aβ peptides. Mutations in these PS genes lead to familial Alzheimer's disease (FAD), providing a basis for study. Tg mice engineered to express familial mutations in either PS1 or PS2 exhibit altered processing of murine APP. However, these mice do not develop plaques, indicating the necessity of hAPP expression to replicate the Aβ deposition seen in AD. The key difference between rodent Aβ and human Aβ lies in three amino acid variations, which affect the aggregation and plaque formation capability of rodent Aβ. To overcome the limitations of single transgenic models, a new generation of double transgenic mice was developed, co-expressing mutations in hAPP and either PS1 or PS2. Since the late 1990s and early 2000s, several APP/PS mutation-based models have been developed for AD research. These models, including knock-in models, are based on various combinations of mutations within the APP gene and PS genes. They utilize distinct promoters and individually exhibit different AD hallmarks as well as varying times of disease manifestation.[153]

1.3.2.2.1 Tg2576 × PS1

The APP/PS1 (Tg2576 × PS1) model was among the first transgenic mouse models developed to incorporate mutations in both the APP and PS1 genes (Table 1.5). This model is generated by interbreeding Tg2576 mice, which overexpress the hAPP gene carrying the Swedish mutation (KM670/671NL), with PS1 mutant mice, which harbor the M146L mutation in the PS1 gene. The M146L mutation, first identified in the mid-1990s, involves the substitution of methionine with leucine at position 146, a change associated with early-onset FAD.[154]

- Pathophysiology and Disease Progression
  - **Aβ Pathology**
    This model exhibits increased Aβ concentrations as early as three to four months of age, initially in the frontal cortex. By 7 months, Aβ accumulation extends to the cerebral cortex, hippocampus, thalamus, and amygdala, and by 12 months, widespread Aβ deposits are present throughout the brain.
  - **Glial Activation and Neuronal Loss**
    Microglial activation surrounding Aβ deposits is observed as early as three months, while clusters of reactive astrocytes appear around plaques by six to seven months.[155] Despite the accelerated Aβ accumulation, significant neuronal loss is not observed in this model.[156]
  - **Cognitive Deficits**
    It has been reported that the APP/PS1 model exhibited progressive age-dependent cognitive impairments across various behavioral tasks. Deficiencies in acquiring skills in the Morris water maze and working memory in the radial arm water maze were apparent between the ages of 5–7 months and 15–17 months, respectively.[157] Additionally, Y-maze performance impairments were observed in mice aged 12–14 weeks.[158] However, in the elevated plus maze test, no significant differences in anxiety or emotionality were detected between transgenic and non-transgenic mice.[157]

1.3.2.2.2 APPswe/PSEN1ΔE9 Model

The APP/PS1 transgenic mouse model carries both the hAPP gene with the Swedish mutation (KM670/671NL) and the PS1 gene with a deletion of exon 9 (Table 1.5). This deletion results from a nonsense mutation at the splice acceptor site, where guanine is replaced by thymine, ultimately preventing the translation of exon 9.[159]

- Pathophysiology and Disease Progression
  - **Neuroinflammation**
    This model exhibits several neuropathological features associated with AD. Elevated levels of inflammatory mediators, along with pronounced astrogliosis and microgliosis, have been reported in both the hippocampus and cortex.[160]
  - **Neurogenesis Deficits and Synaptic Dysfunction**
    It has been observed that eight-month-old APPswe/PSEN1ΔE9 mice display a significant reduction in the number of immature neurons and neural progenitors within the dentate gyrus compared to age-matched wild-type controls.[153]
    Neurotrophic factor dysregulation has also been documented in this model. Specifically, decreased levels of BDNF and NGF have been detected in the hippocampus of seven-month-old APPswe/PSEN1ΔE9 mice.[161]
    In parallel, this model exhibits synaptic dysfunction, characterized by reduced levels of key synaptic proteins, including synaptophysin, postsynaptic density(PSD)-93, and PSD-95, mirroring synaptic deficits commonly observed in AD.[161]

1.3.2.2.3 5XFAD

In 2006, a new transgenic line was developed with five mutations. This model carries the double Swedish mutation (K670N/M671L) in the APP gene, which enhances amyloid production, along with the Florida (I716V) and London (V717I) mutations in APP that further promote amyloidogenic processing. Additionally, two PS1 mutations (M146L and L286V) contribute to altered γ-secretase activity, leading to a higher ratio of the more amyloidogenic Aβ species (Table 1.5).[162]

- Pathophysiology and Disease Progression
  - **Aβ Pathology**
    The 5XFAD model exhibits rapid Alzheimer's-like pathology, characterized by extensive amyloid plaque deposition at an earlier stage than other transgenic models. However, its limited phosphorylated tau pathology makes it less suitable for studying tauopathy-related aspects of the disease.[163] Intracellular Aβ accumulation begins as early as 1.5 months of age, with extracellular deposition observed by 2 months. Cognitive deficits in hippocampus-dependent tasks emerge between four and six months, coinciding with moderate Aβ accumulation and hippocampal synaptic dysfunction in the Schaffer collateral-CA1 pathway.[164] Additionally, BACE1 levels increase approximately two-fold in the brains of 5XFAD mice as amyloid pathology progresses.[165,166]
  - **Cognitive Impairments**
    Interestingly, early cognitive impairments in the frontal cortex appear by 4 months, preceding hippocampal-dependent deficits, and are associated with neuropathological changes such as gliosis and amyloid plaque formation.[167]
  - **Synaptic Dysfunction and Neuronal Loss**
    At five to six months, presynaptic dystrophic neurites are detected in regions surrounding amyloid plaques.[168] Selective neuronal loss has been reported in this model, particularly affecting noradrenergic[169] and cholinergic populations.[170]
  - **Neuroinflammation**
    Neuroinflammation is also reported in the 5XFAD model, characterized by the activation of pro-inflammatory cytokines and immune markers in the cortex and hippocampus. This inflammatory response contributes to apoptosis-induced neurodegeneration, primarily mediated by microglial activation and caspase-3 signaling.[171]
  - **Non-Cognitive Impairments**
    Beyond cognitive impairments, 5XFAD mice exhibit additional non-cognitive abnormalities, including age-related motor dysfunction, working memory deficits, and reduced anxiety levels. Moreover, in this model, neurodegenerative changes are not limited to the brain; substantial Aβ deposition and plaque formation occur in the spinal cord, accompanied by age-dependent axonopathy.[172]
  - **Circadian Rhythm Disruptions**
    Disruptions in sleep–wake cycles resembling those observed in AD patients have also been reported in 5XFAD mice.[173]

    These findings highlight the utility of this model in studying both the neuropathological and behavioral aspects of AD.

1.3.2.3 Tau Mutation-Based Mice

1.3.2.3.1 P301S Transgenic Mouse Model

The P301S mouse model expresses a mutant (P301S) human microtubule-associated protein tau (MAPT) gene under the mouse prion-protein promoter (Prnp) on a B6C3H/F1 background (Table 1.5).[174] Age-dependent tau aggregation begins at three months, with NFTs forming by six months. Severe neuron loss and hippocampal/entorhinal atrophy occur in 9–12 months.[174]

- Pathophysiology and Disease Progression
  - Early microglial activation precedes tangle formation.
  - Synaptic dysfunction, evident by three months, preceding tau aggregation, with cortical and hippocampal atrophy mirroring human tauopathy progression.[174]

- Cognitive impairments, apparent as early as three months of age, with significant impairments in social cognition and social preference emerging between five and six months.[175]
- Progressive neurogenic sarcopenia reflecting peripheral atrophy.[176]
- Visual impairments, evident at 6 months, attributed in part to the dysfunction of glutamatergic synapses of the retina.[177]
- Sleep–wake cycle disruptions showed by hyperarousal with an early onset and progressive nature, along with specific disruption of dark-phase sleep, noted in females at three months and in males at six months.[178]
- Sexual dimorphism in behavior, plasma proteins, and neuropathology; male mice preferred for anti-phosphorylated tau therapy testing.[179]
- Structural changes in the Golgi apparatus of pyramidal neurons.[180]
- AMPA receptor alterations on synaptic and extrasynaptic membranes of various neuron types, impairing synaptic transmission.[181]
- Impaired mitochondrial fission and fusion.[182]

1.3.2.3.2 rTg4510 Model

The rTg4510 mouse model is characterized by the overexpression of the mutant P301L human MAPT gene. This genetic modification leads to the progressive accumulation of pathological tau, a hallmark of tauopathies. Tauopathy in the rTg4510 line emerges in 4–5 months and progresses to a severe phase by 8–12 months.[183]

- Pathophysiology and Disease Progression
  - Synaptic Loss and brain atrophy.[184]
  - Neuroinflammation linked to tau accumulation.[185]
  - Rapid spatial learning and memory decline in rTg4510 mice, associated with increased hyperphosphorylated tau accumulation between 3 and 5.5 months.[186]
  - Greater accumulation of pathological tau in the brain and elevated total tau in CSF in female rTg4510 mice compared to males.[187]
  - Sex-dependent tau pathology and memory impairment, with earlier onset and greater severity in females, marked by elevated tau levels and more pronounced cognitive impairments.[186]
  - Disruption of short- and long-term visual plasticity at 5 months (early stages of tauopathy), progressively worsening with age, accompanied by impaired visually evoked behaviors.[188]
  - Altered mitochondrial trafficking in neurons.[176]
  - Disrupted glymphatic cerebrospinal fluid–interstitial fluid (CSF–ISF) exchange, affecting the clearance of protein species like Aβ and tau, along with altered aquaporin-4 (AQP4) polarization.[189]
  - Disturbances in sleep/wake cycles.[190]

1.3.2.4 Amyloid Precursor Protein, Presenilins, and Tau Mutation-Based Mice

1.3.2.4.1 3×Tg-AD Model

In 2003, researchers developed the triple-transgenic AD (3×Tg-AD) mouse model to investigate the interaction between Aβ and tau proteins and their effects on synaptic function. This model carries three transgenes: presenilin-1 (PS1M146V), APP (APPSwe), and tauP301L (Table 1.5), all of which are associated with AD pathology. By incorporating these transgenes into a single genetic background, researchers aimed to replicate the complex interplay between Aβ, tau, and synaptic dysfunction in an animal model of FAD.

To generate 3×Tg-AD mice, researchers used a microinjection technique instead of crossing independent transgenic lines. In this method, two transgenes (APPSwe and tauP301L) were introduced into single-cell embryos from homozygous PS1M146V knock-in mice. This approach ensured a uniform genetic background, facilitating controlled investigations of AD-related pathologies.[191]

- Pathophysiology and Disease Progression
  - **Aβ Pathology**
    The 3×Tg-AD mice exhibit a progressive and region-specific development of Aβ and tau pathology, closely mirroring human AD. Aβ deposition begins in the cortex before spreading to the hippocampus, while tau pathology originates in the hippocampus and later extends to the cortex. Despite the similar overexpression levels of hAPP and tau transgenes, Aβ accumulation precedes the formation of tau tangles, supporting the amyloid cascade hypothesis, which posits that Aβ aggregation is an early pathological event triggering downstream tau pathology in AD.
  - **Neuroinflammation and Glial Responses**
    - Aβ deposition is detectable by three months of age, with increased abundance by six months.
    - At six months, the entorhinal cortex exhibits elevated TNF-α and MCP-1 mRNA levels prior to amyloid plaque formation.[192]
    - Microglial proliferation and activation correlate with the onset of fibrillar Aβ plaques and tau hyperphosphorylation.[193]
    - Astroglial atrophy is widespread, though astrocytes surrounding plaques exhibit hypertrophy.[194]
    - A decline in glutamine synthetase-immunoreactive (GS-IR) astrocytes is observed from 12 to 18 months, particularly in the dentate gyrus and CA1 region.[195]
  - **Synaptic and Neurotransmitter Alterations**
    - Age-dependent reductions in nicotinic acetylcholine receptor alpha are evident by six months and correlate with intraneuronal Aβ42 accumulation.[196]
    - M1 muscarinic receptors regulate AD-like symptoms in 3×Tg-AD mice.[197]
    - The medial septum-dentate gyrus GABAergic circuit undergoes remodeling as a compensatory response to impaired GABAergic transmission.[198]
  - **Sex Differences in AD Pathology**
    - Cognitive deficits in 3×Tg-AD mice become sexually dimorphic after four months. Female mice exhibit greater impairments in stress-related tasks due to heightened corticosterone responses.[199]
    - Female mice develop more aggressive Aβ pathology, potentially due to increased beta-secretase activity and decreased neprilysin levels, leading to higher Aβ production and reduced degradation.[200]
    - No gender differences in the levels of phosphorylated tau.[200]
  - **Neurogenesis and Early Developmental Defects**
    - Neurogenesis in the dentate gyrus is impaired as early as six months, with female mice showing deficits as early as four months.[201]
    - Neural stem/progenitor cell populations are significantly reduced before AD pathology manifests, with RNA-seq revealing early gene expression changes in Notch and Wnt signaling pathways.[202]
  - **Mitochondrial Dysfunction and Oxidative Stress**
    - Mitochondrial respiration is reduced in female 3×Tg-AD mice by three months, accompanied by decreased levels and activity of pyruvate dehydrogenase (PDH), increased oxidative stress, and lipid peroxidation.[203]
    - Mitochondrial Aβ accumulation peaks at 9 months and correlates with increased binding to ABAD, a protein involved in alcohol metabolism.[203]

- Structural mitochondrial changes, including reduced numeric density and increased volume, occur in the CA1 pyramidal neurons.[204]
- Synaptic mitochondrial dynamics and quality control mechanisms are impaired with aging.[205]
- Astrocytic bioenergetics, proteostasis, and Ca^{2+} homeostasis are deregulated due to disrupted mitochondria–ER interactions.[205]

- **Synaptic Plasticity and Behavioral Deficits**
  - At 13 months, a 28.5% reduction in perforated synapses is noted, indicating compromised synaptic plasticity.[206]
  - 3×Tg-AD mice display heightened fear responses alongside Aβ accumulation in GABAergic and glutamatergic neurons within the basolateral amygdala.[207]
  - Altered myelination patterns in specific brain regions precede amyloid and tau pathology.[208]
- **Vascular Dysfunctions**
  - Plasma and cerebrospinal fluid Aβ1–42 levels show an inverse correlation until ThS-positive plaques form.[209]
  - A significant region-dependent reduction in cerebrovascular volume, particularly in the hippocampus, is observed at 11 months.[210]
  - Neurovascular coupling is impaired before cognitive decline, marked by reduced cerebral blood flow responsiveness to NO signals.[211]
  - Vascular remodeling occurs in four to six months, coinciding with cognitive impairment onset.[212]
- **Circadian Rhythm Disruptions**
  - Circadian rhythm disturbances manifest at six months and are more pronounced in males, with altered Per1 and Per2 gene expression in the suprachiasmatic nucleus.[213]
  - Male 3×Tg-AD mice exhibit increased daytime movement and shorter free-running cycles, while females show reduced nocturnal activity.[214]
  - The orexin system and core clock gene expression differ significantly from wild-type mice, with correlations to AD risk genes.[215]
- **Dopaminergic Alterations**
  - Early dopaminergic dysfunction with increased D2/3 receptor density in the striatum and midbrain, leading to altered anxiety-like behavior, increased locomotion, and heightened receptor sensitivity.[216]
- **Gut Microbiota Alterations**
  - Dysbiosis of gut microbiota is evident by three months, with altered microbial composition differing between male and female 3×Tg-AD mice.[216]

REFERENCES

1. Chen MK, Mecca AP, Naganawa M, Finnema SJ, Toyonaga T, Lin SF, et al. Assessing synaptic density in Alzheimer disease with synaptic vesicle glycoprotein 2A positron emission tomographic imaging. JAMA Neurology. 2018 Oct;75(10):1215–24.
2. Monzio Compagnoni G, Di Fonzo A, Corti S, Comi GP, Bresolin N, Masliah E. The role of mitochondria in neurodegenerative diseases: The lesson from Alzheimer's disease and Parkinson's disease. Molecular Neurobiology. 2020 Jul;57(7):2959–80.
3. Kisler K, Nelson AR, Montagne A, Zlokovic B V. Cerebral blood flow regulation and neurovascular dysfunction in Alzheimer disease. Nature Reviews Neuroscience. 2017 Jul;18(7):419–34.
4. Jellinger KA. Neuropathological assessment of the Alzheimer spectrum. Journal of Neural Transmission (Vienna, Austria:1996). 2020 Sep;127(9):1229–56.
5. Sochocka M, Donskow-Łysoniewska K, Diniz BS, Kurpas D, Brzozowska E, Leszek J. The gut microbiome alterations and inflammation-driven pathogenesis of Alzheimer's disease – A critical review. Molecular Neurobiology. 2019 Mar;56(3):1841–51.

6. Scheltens P, De Strooper B, Kivipelto M, Holstege H, Chételat G, Teunissen CE, et al. Alzheimer's disease. Lancet (London, England). 2021 Apr;397(10284):1577–90.
7. Peter-Derex L, Yammine P, Bastuji H, Croisile B. Sleep and Alzheimer's disease. Sleep Medicine Reviews. 2015 Feb;19:29–38.
8. Diekelmann S, Born J. The memory function of sleep. Nature Reviews Neuroscience. 2010 Feb;11(2):114–26.
9. Buckley RF, Mormino EC, Rabin JS, Hohman TJ, Landau S, Hanseeuw BJ, et al. Sex differences in the association of global amyloid and regional tau deposition measured by positron emission tomography in clinically normal older adults. JAMA Neurology. 2019 May;76(5):542–51.
10. Klatzo I, Wisniewski H, Streicher E. Experimental production of neurofibrillary degeneration: I. Light microscopic observations. Journal of Neuropathology & Experimental Neurology [Internet]. 1965 Apr 1 [cited 2024 May 5];24(2):187–99. Available from: https://dx.doi.org/10.1097/00005072-196504000-00002
11. Yumoto S, Nagai H, Matsuzaki H, Matsumura H, Tada W, Nagatsuma E, et al. Aluminium incorporation into the brain of rat fetuses and sucklings. Brain Research Bulletin. 2001 May 15;55(2):229–34.
12. Uemura E. Intranuclear aluminum accumulation in chronic animals with experimental neurofibrillary changes. Experimental Neurology [Internet]. 1984 Jul 1 [cited 2024 May 5];85(1):10–8. Available from: https://europepmc.org/article/med/6734771
13. Savory J, Herman MM, Ghribi O. Intracellular mechanisms underlying aluminum-induced apoptosis in rabbit brain. Journal of Inorganic Biochemistry. 2003 Sep 15;97(1):151–4.
14. Song J. Animal model of aluminum-induced Alzheimer's disease. Advances in Experimental Medicine and Biology [Internet]. 2018 [cited 2024 May 5];1091:113–27. Available from: https://link.springer.com/chapter/10.1007/978-981-13-1370-7_7
15. Thawkar BS, Kaur G. Betanin combined with virgin coconut oil inhibits neuroinflammation in aluminum chloride-induced toxicity in rats by regulating NLRP3 inflammasome. Journal of Traditional and Complementary Medicine. 2024 May 1;14(3):287–99.
16. Gao J, Liu W, Liu J, Hao N, Pei J, Zhang L. The role of acetylation and methylation of rat hippocampal histone H3 in the mechanism of aluminum-induced neurotoxicity. Neurochemical Research [Internet]. 2024 Feb 1 [cited 2024 May 9];49(2):441–52. Available from: https://pubmed.ncbi.nlm.nih.gov/37897558/
17. Geravand S, Karami M, Sahraei H, Rahimi F. Protective effects of L-arginine on Alzheimer's disease: Modulating hippocampal nitric oxide levels and memory deficits in aluminum chloride-induced rat model. European Journal of Pharmacology. 2023 Nov 5;958:176030.
18. Liaquat L, Ahmad S, Sadir S, Batool Z, Khaliq S, Tabassum S, et al. Development of AD like symptoms following co-administration of AlCl3 and D-gal in rats: A neurochemical, biochemical and behavioural study. Pakistan Journal of Pharmaceutical Sciences [Internet]. 2017;30 (2(Suppl)):647–53. Available from: https://api.semanticscholar.org/CorpusID:22776131
19. AbdEl-Raouf K, Farrag HSH, Rashed R, Ismail MA, El-Ganzuri MA, El-Sayed WM. New bithiophene derivative attenuated Alzheimer's disease induced by aluminum in a rat model via antioxidant activity and restoration of neuronal and synaptic transmission. Journal of Trace Elements in Medicine and Biology. 2024 Mar 1;82:127352.
20. Liu W, Liu J, Gao J, Duan X, Zhang L. Effects of subchronic aluminum exposure on learning, memory, and neurotrophic factors in rats. Neurotoxicity Research [Internet]. 2022 Dec 1 [cited 2024 May 10];40(6):2046–60. Available from: https://link.springer.com/article/10.1007/s12640-022-00599-z
21. Jabeen K, Rehman K, Akash MSH, Nadeem A, Mir TM. Neuroprotective and cardiometabolic role of Vitamin E: Alleviating neuroinflammation and metabolic disturbance induced by AlCl3 in rat models. Biomedicines 2023;11:2453 [Internet]. 2023 Sep 4 [cited 2024 May 9];11(9):2453. Available from: https://www.mdpi.com/2227-9059/11/9/2453/htm
22. Sentyabreva AV, Miroshnichenko EA, Melnikova EA, Tsvetkov IS, Kosyreva AM. Morphofunctional changes in brain and peripheral blood in adult and aged Wistar Rats with AlCl3-induced neurodegeneration. Biomedicines 2023;11:2336 [Internet]. 2023 Aug 22 [cited 2024 May 9];11(9):2336. Available from: https://www.mdpi.com/2227-9059/11/9/2336/htm
23. Li C, Wang N, Zheng G, Yang L. Oral administration of resveratrol-selenium-peptide nanocomposites alleviates Alzheimer's disease-like pathogenesis by inhibiting Aβ aggregation and regulating gut microbiota. ACS Applied Materials & Interfaces [Internet]. 2021 Oct 6 [cited 2024 May 10];13(39):46406–20. Available from: https://pubmed.ncbi.nlm.nih.gov/34569225/
24. Liu J, Ye T, Zhang Y, Zhang R, Kong Y, Zhang Y, et al. Protective effect of ginkgolide B against cognitive impairment in mice via regulation of gut microbiota. Journal of Agricultural and Food Chemistry [Internet]. 2021 Oct 20 [cited 2024 May 10];69(41):12230–40. Available from: https://pubmed.ncbi.nlm.nih.gov/34633804/

25. Hassan SA, Kadry MO. Neurodegenerative and hepatorenal disorders induced via aluminum chloride in murine system: Impact of β-secretase, MAPK, and KIM. Biological Trace Element Research [Internet]. 2021 Jan 1 [cited 2024 May 10];199(1):227–36. Available from: https://pubmed.ncbi.nlm.nih.gov/32279189/
26. Chen GF, Xu TH, Yan Y, Zhou YR, Jiang Y, Melcher K, et al. Amyloid beta: Structure, biology and structure-based therapeutic development. Nature Publishing Group [Internet]. 2017 [cited 2024 Mar 27];38:1205–35. Available from: www.chinaphar.com
27. Yankner BA, Duffy LK, Kirschner DA. Neurotrophic and neurotoxic effects of amyloid beta protein: Reversal by tachykinin neuropeptides. Science (New York, NY) [Internet]. 1990 [cited 2024 Mar 27];250(4978):279–82. Available from: https://pubmed.ncbi.nlm.nih.gov/2218531/
28. Frautschy SA, Baird A, Cole GM. Effects of injected Alzheimer β-amyloid cores in rat brain. Proceedings of the National Academy of Sciences of the United States of America [Internet]. 1991 Oct 1 [cited 2024 Mar 27];88(19):8362–6. Available from: https://click.endnote.com/viewer?doi=10.1073%2Fpnas.88.19.8362&token=WzMxMjcwNiwiMTAuMTA3My9wbmFzLjg4LjE5LjgzNjIiXQ.vz3qyWQ0UgcGjRGKuuyotQ1KkYQ
29. Epelbaum S, Youssef I, Lacor PN, Chaurand P, Duplus E, Brugg B, et al. Acute amnestic encephalopathy in amyloid-β oligomer-injected mice is due to their widespread diffusion in vivo. Neurobiology of Aging [Internet]. 2015 Jun 1 [cited 2024 Apr 15];36(6):2043–52. Available from: https://pubmed.ncbi.nlm.nih.gov/25862419/
30. Calvo-Flores Guzmán B, Elizabeth Chaffey T, Hansika Palpagama T, Waters S, Boix J, Tate WP, et al. The interplay between beta-amyloid 1–42 ($A\beta_{1-42}$)-induced hippocampal inflammatory response, p-tau, vascular pathology, and their synergistic contributions to neuronal death and behavioral deficits. Frontiers in Molecular Neuroscience. 2020 Nov 2;13:552073.
31. Chen C, Li X, Gao P, Tu Y, Zhao M, Li J, et al. Baicalin attenuates Alzheimer-like pathological changes and memory deficits induced by amyloid β1-42 protein. Metabolic Brain Disease [Internet]. 2015 Apr 1 [cited 2024 Apr 15];30(2):537–44. Available from: https://pubmed.ncbi.nlm.nih.gov/25108596/
32. Xu S, Guan Q, Wang C, Wei X, Chen X, Zheng B, et al. Rosiglitazone prevents the memory deficits induced by amyloid-beta oligomers via inhibition of inflammatory responses. Neuroscience Letters [Internet]. 2014 Aug 22 [cited 2024 Apr 15];578:7–11. Available from: https://pubmed.ncbi.nlm.nih.gov/24933538/
33. Joy Nell H, Whitehead SN, Cechetto DF. Age-dependent effect of β-amyloid toxicity on basal forebrain cholinergic neurons and inflammation in the rat brain. 2014 [cited 2024 Mar 27]; Available from: https://onlinelibrary.wiley.com/doi/10.1111/bpa.12199
34. Wong RS, Cechetto DF, Whitehead SN. Assessing the effects of acute amyloid β oligomer exposure in the rat. International Journal of Molecular Sciences. 2016 Sep 1;17(9).
35. Brkic M, Balusu S, Van Wonterghem E, Gorlé N, Benilova I, Kremer A, et al. Amyloid β oligomers disrupt blood–CSF barrier integrity by activating matrix metalloproteinases. Journal of Neuroscience [Internet]. 2015 Sep 16 [cited 2024 Apr 15];35(37):12766–78. Available from: https://www.jneurosci.org/content/35/37/12766
36. Balducci C, Frasca A, Zotti M, La Vitola P, Mhillaj E, Grigoli E, et al. Toll-like receptor 4-dependent glial cell activation mediates the impairment in memory establishment induced by β-amyloid oligomers in an acute mouse model of Alzheimer's disease. Brain, Behavior, and Immunity [Internet]. 2017 Feb 1 [cited 2024 Apr 15];60:188–97. Available from: https://pubmed.ncbi.nlm.nih.gov/27751869/
37. Gupta P, Sil S, Ghosh R, Ghosh A, Ghosh T. Intracerebroventricular Aβ-induced neuroinflammation alters peripheral immune responses in rats. Journal of Molecular Neuroscience [Internet]. 2018 Dec 1 [cited 2024 Apr 15];66(4):572–86. Available from: https://link.springer.com/article/10.1007/s12031-018-1189-9
38. Tang SS, Hong H, Chen L, Mei Z-L, Ji M-J, Xiang G-Q, et al. Involvement of cysteinyl leukotriene receptor 1 in Aβ1-42-induced neurotoxicity in vitro and in vivo. Neurobiology of Aging [Internet]. 2014 Mar [cited 2024 Apr 15];35(3):590–9. Available from: https://pubmed.ncbi.nlm.nih.gov/24269024/
39. Stepanichev M, Lazareva N, Tukhbatova G, Salozhin S, Gulyaeva N. Transient disturbances in contextual fear memory induced by Aβ(25–35) in rats are accompanied by cholinergic dysfunction. Behavioural Brain Research. 2014 Feb 1;259:152–7.
40. Shaerzadeh F, Motamedi F, Minai-Tehrani D, Khodagholi F. Monitoring of neuronal loss in the hippocampus of Aβ-injected rat: Autophagy, mitophagy, and mitochondrial biogenesis stand against apoptosis. Neuromolecular Medicine [Internet]. 2014 Mar [cited 2024 Apr 15];16(1):175–90. Available from: https://pubmed.ncbi.nlm.nih.gov/24203394/

41. Xu Y, Cao DH, Wu GM, Hou XY. Involvement of P38MAPK activation by NMDA receptors and non-NMDA receptors in amyloid-peptide-induced neuronal loss in rat hippocampal CA1 and CA3 subfields. Neuroscience Research [Internet]. 2014 [cited 2024 Mar 28];85:51–7. Available from: http://dx.doi.org/10.1016/j.neures.2014.05.011
42. Tucci P, Mhillaj E, Morgese MG, Colaianna M, Zotti M, Schiavone S, et al. Memantine prevents memory consolidation failure induced by soluble beta amyloid in rats. Frontiers in Behavioral Neuroscience. 2014 Sep 19;8.
43. Mayordomo-Cava J, Iborra-Lázaro G, Djebari S, Temprano-Carazo S, Sánchez-Rodríguez I, Jeremic D, et al. Impairments of synaptic plasticity induction threshold and network oscillatory activity in the hippocampus underlie memory deficits in a non-transgenic mouse model of amyloidosis. Biology [Internet]. 2020;9(7). Available from: https://www.mdpi.com/2079-7737/9/7/175
44. Yeung JHY, Palpagama TH, Tate WP, Peppercorn K, Waldvogel HJ, Faull RLM, et al. The acute effects of amyloid-beta1–42 on glutamatergic receptor and transporter expression in the mouse hippocampus. Frontiers in Neuroscience [Internet]. 2020 Jan 17 [cited 2024 Apr 15];13:494393. Available from: www.frontiersin.org
45. Liu T, Bai W, Yi H, Tan T, Wei J, Wang J, et al. Functional connectivity in a rat model of Alzheimer's disease during a working memory task. Current Alzheimer Research. 2014 Dec 9;11(10):981–91.
46. Bai W, Yi H, Liu T, Wei J, Tian X. Incoordination between spikes and LFPs in Aβ1-42-mediated memory deficits in rats. Frontiers in Behavioral Neuroscience [Internet]. 2014 Nov 27 [cited 2024 Apr 15];8(NOV). Available from: https://pubmed.ncbi.nlm.nih.gov/25505877/
47. Ghasemi R, Zarifkar A, Rastegar K, Maghsoudi N, Moosavi M. Repeated intra-hippocampal injection of beta-amyloid 25–35 induces a reproducible impairment of learning and memory: Considering caspase-3 and MAPKs activity. European Journal of Pharmacology [Internet]. 2014 Mar 5 [cited 2024 Apr 15];726(1):33–40. Available from: https://pubmed.ncbi.nlm.nih.gov/24418687/
48. Klyubin I, Wang Q, Reed MN, Irving EA, Upton N, Hofmeister J, et al. Protection against Aβ-mediated rapid disruption of synaptic plasticity and memory by memantine. Neurobiology of Aging [Internet]. 2011 Apr [cited 2024 Apr 15];32(4):614–23. Available from: https://pubmed.ncbi.nlm.nih.gov/19446369/
49. Sánchez-Rodríguez I, Temprano-Carazo S, Nájera A, Djebari S, Yajeya J, Gruart A, et al. Publisher Correction: Activation of G-protein-gated inwardly rectifying potassium (Kir3/GirK) channels rescues hippocampal functions in a mouse model of early amyloid-β pathology. Scientific Reports [Internet]. 2020 Jan 29 [cited 2024 Apr 15];10(1):1722. Available from: https://pubmed.ncbi.nlm.nih.gov/31996714/
50. Wang BY, Zhong Y, Zhao Z, Miao Y. Epigenetic suppression of hippocampal BDNF mediates the memory deficiency induced by amyloid fibrils. Pharmacology, Biochemistry, and Behavior [Internet]. 2014 [cited 2024 Apr 15];126:83–9. Available from: https://pubmed.ncbi.nlm.nih.gov/25242807/
51. Canas PM, Simões AP, Rodrigues RJ, Cunha RA. Predominant loss of glutamatergic terminal markers in a β-amyloid peptide model of Alzheimer's disease. Neuropharmacology. 2014 Jan 1;76(PART A):51–6.
52. Yeung JHY, Calvo-Flores Guzmán B, Palpagama TH, Ethiraj J, Zhai Y, Tate WP, et al. Amyloid-beta1-42 induced glutamatergic receptor and transporter expression changes in the mouse hippocampus. Journal of Neurochemistry [Internet]. 2020 Oct 1 [cited 2024 Apr 15];155(1):62–80. Available from: https://pubmed.ncbi.nlm.nih.gov/32491248/
53. Bie B, Wu J, Foss JF, Naguib M. Amyloid fibrils induce dysfunction of hippocampal glutamatergic silent synapses. Hippocampus [Internet]. 2018 May 7 [cited 2024 Apr 15];28(8):549–56. Available from: https://europepmc.org/articles/PMC6133714
54. Detrait E, Maurice T, Hanon E, Leclercq K, Lamberty Y. Lack of synaptic vesicle protein SV2B protects against amyloid-25-35-induced oxidative stress, cholinergic deficit and cognitive impairment in mice. Behavioural Brain Research [Internet]. 2014 [cited 2024 Mar 28];271:277–85. Available from: http://dx.doi.org/10.1016/j.bbr.2014.06.013
55. Morgese MG, Colaianna M, Mhillaj E, Zotti M, Schiavone S, D'Antonio P, et al. Soluble beta amyloid evokes alteration in brain norepinephrine levels: role of nitric oxide and interleukin-1. Frontiers in Neuroscience [Internet]. 2015 [cited 2024 Apr 15];9(NOV). Available from: https://pubmed.ncbi.nlm.nih.gov/26594145/
56. Canet G, Pineau F, Zussy C, Hernandez C, Hunt H, Chevallier N, et al. Glucocorticoid receptors signaling impairment potentiates amyloid-β oligomers-induced pathology in an acute model of Alzheimer's disease. FASEB Journal : Official Publication of the Federation of American Societies for Experimental Biology [Internet]. 2020 Jan 1 [cited 2024 Apr 15];34(1):1150–68. Available from: https://pubmed.ncbi.nlm.nih.gov/31914623/
57. Morgese MG, Schiavone S, Mhillaj E, Bove M, Tucci P, Trabace L. N-3 PUFA diet enrichment prevents amyloid beta-induced depressive-like phenotype. Pharmacological Research. 2018 Mar 1;129:526–34.

58. James D, Kang S, Park S. Injection of β-amyloid into the hippocampus induces metabolic disturbances and involuntary weight loss which may be early indicators of Alzheimer's disease. Aging Clinical and Experimental Research. 2014;26(1):93–8.
59. Clarke JR, Lyra e Silva NM, Figueiredo CP, Frozza RL, Ledo JH, Beckman D, et al. Alzheimer-associated Aβ oligomers impact the central nervous system to induce peripheral metabolic deregulation. EMBO Molecular Medicine [Internet]. 2015 Feb [cited 2024 Apr 16];7(2):190–210. Available from: https://pubmed.ncbi.nlm.nih.gov/25617315/
60. Matsuzaki K, Katakura M, Sugimoto N, Hara T, Hashimoto M, Shido O. β-amyloid infusion into lateral ventricle alters behavioral thermoregulation and attenuates acquired heat tolerance in rats. Temperature (Austin, Tex) [Internet]. 2015 Jul 3 [cited 2024 Apr 16];2(3):418–24. Available from: https://pubmed.ncbi.nlm.nih.gov/27227055/
61. De Oliveira J, Moreira ELG, Dos Santos DB, Piermartiri TC, Dutra RC, Pinton S, et al. Increased susceptibility to amyloid-β-induced neurotoxicity in mice lacking the low-density lipoprotein receptor. Journal of Alzheimer's Disease. 2014 Jan 1;41(1):43–60.
62. Ledezma C, Coria-Lucero C, Delsouc MB, Casais M, Della Vedova C, Ramirez D, et al. Effect of an intracerebroventricular injection of aggregated beta-Amyloid (1-42) on daily rhythms of oxidative stress parameters in the prefrontal cortex. Neuroscience [Internet]. 2021 Mar 15 [cited 2024 Apr 16]; 458:99–107. Available from: https://pubmed.ncbi.nlm.nih.gov/32827572/
63. Takeda A, Nakamura M, Fujii H, Uematsu C, Minamino T, Adlard PA, et al. Amyloid b-mediated Zn 2+ influx into dentate granule cells transiently induces a short-term cognitive deficit. 2014.
64. Baleriola J, Walker CA, Jean YY, Crary JF, Troy CM, Nagy PL, et al. Axonally synthesized ATF4 transmits a neurodegenerative signal across brain regions. Cell [Internet]. 2014 Aug 28 [cited 2024 Apr 16];158(5):1159–72. Available from: https://pubmed.ncbi.nlm.nih.gov/25171414/
65. Choi YY, Maeda T, Fujii H, Yokozawa T, Kim HY, Cho EJ, et al. Oligonol improves memory and cognition under an amyloid β 25-35-induced Alzheimer's mouse model. 2014 [cited 2024 Apr 3]; Available from: http://dx.doi.org/10.1016/j.nutres.2014.06.008
66. Naderi S, Khodagholi F, Pourbadie HG, Naderi N, Rafiei S, Janahmadi M, et al. Role of amyloid beta (25-35) neurotoxicity in the ferroptosis and necroptosis as modalities of regulated cell death in Alzheimer's disease. Neurotoxicology [Internet]. 2023 Jan 1 [cited 2024 Apr 16];94:71–86. Available from: https://pubmed.ncbi.nlm.nih.gov/36347329/
67. He B, Zheng M, Liu Q, Shi Z, Long S, Lu X, et al. Injected amyloid beta in the olfactory bulb transfers to other brain regions via neural connections in mice. Molecular Neurobiology [Internet]. 2018 Feb 1 [cited 2024 Apr 16];55(2):1703–13. Available from: https://pubmed.ncbi.nlm.nih.gov/28211008/
68. Takeda A, Tamano H, Tempaku M, Sasaki M, Uematsu C, Sato S, et al. Extracellular Zn2+ is essential for amyloid β1-42-induced cognitive decline in the normal brain and its rescue. The Journal of Neuroscience : The Official Journal of the Society for Neuroscience [Internet]. 2017 Jul 26 [cited 2024 Apr 16];37(30):7253–62. Available from: https://pubmed.ncbi.nlm.nih.gov/28652412/
69. Tamano H, Takiguchi M, Tanaka Y, Murakami T, Adlard PA, Bush AI, et al. Preferential neurodegeneration in the dentate gyrus by amyloid β1-42-induced intracellular Zn^{2+}dysregulation and its defense strategy. Molecular Neurobiology [Internet]. 2020 Apr 1 [cited 2024 Apr 16];57(4):1875–88. Available from: https://pubmed.ncbi.nlm.nih.gov/31865526/
70. Tamano H, Suzuki H, Kobuchi S, Adlard PA, Bush AI, Takeda A. Difference in ability for extracellular Zn2+ influx between human and rat amyloid β1-42 and its significance. Neurotoxicology [Internet]. 2019 May 1 [cited 2024 Apr 16];72:1–5. Available from: https://pubmed.ncbi.nlm.nih.gov/30664975/
71. Cho SM, Kim HV, Lee S, Kim HY, Kim W, Kim TS, et al. Correlations of amyloid-β concentrations between CSF and plasma in acute Alzheimer mouse model. Scientific Reports [Internet]. 2014 Oct 27 [cited 2024 Apr 16];4. Available from: https://pubmed.ncbi.nlm.nih.gov/25345439/
72. Fritschi SK, Cintron A, Ye L, Mahler J, Bühler A, Baumann F, et al. Aβ seeds resist inactivation by formaldehyde. Acta Neuropathologica. 2014 Oct 1;128(4):477–84.
73. Yenkoyan K, Margaryan T, Matinyan S, Chavushyan V, Danielyan M, Davtyan T, et al. Effects of β-amyloid (1-42) Administration on the main neurogenic niches of the adult brain: Amyloid-induced neurodegeneration influences neurogenesis. International Journal of Molecular Sciences. 2022 Dec 1;23(23).
74. Goswami P, Afjal MA, Akhter J, Mangla A, Khan J, Parvez S, et al. Involvement of endoplasmic reticulum stress in amyloid β (1-42)-induced Alzheimer's like neuropathological process in rat brain. Brain Research Bulletin. 2020 Dec 1;165:108–17.
75. Aslan Karakelle N, Dinçer S, Yar Sağlam AS. The effect of intracerebroventricular amyloid beta 1-42 application on cognitive functions in aged rats supplemented with taurine and the change of peroxisomal

proteins in this process. Brain Research Bulletin [Internet]. 2021 Jul 1 [cited 2024 Apr 16];172:89–97. Available from: https://pubmed.ncbi.nlm.nih.gov/33892084/

76. Qian X hang, Liu X li, Chen G, Chen S di, Tang H dong. Injection of amyloid-β to lateral ventricle induces gut microbiota dysbiosis in association with inhibition of cholinergic anti-inflammatory pathways in Alzheimer's disease. Journal of neuroinflammation [Internet]. 2022 Dec 1 [cited 2024 Apr 16]; 19(1). Available from: https://pubmed.ncbi.nlm.nih.gov/36171620/
77. Nicole O, Hadzibegovic S, Gajda J, Bontempi B, Bem T, Meyrand P. Soluble amyloid beta oligomers block the learning-induced increase in hippocampal sharp wave-ripple rate and impair spatial memory formation. Scientific Reports 2016 6:1 [Internet]. 2016 Mar 7 [cited 2024 Apr 16];6(1):1–12. Available from: https://www.nature.com/articles/srep22728
78. Navigatore Fonzo L, Alfaro M, Mazaferro P, Golini R, Jorge L, Cecilia Della Vedova M, et al. An intracerebroventricular injection of amyloid-beta peptide (1-42) aggregates modifies daily temporal organization of clock factors expression, protein carbonyls and antioxidant enzymes in the rat hippocampus. Brain Research [Internet]. 2021 Sep 15 [cited 2024 Apr 16];1767. Available from: https://pubmed.ncbi.nlm.nih.gov/33771518/
79. Khakpai F, Nasehi M, Haeri-Rohani A, Eidi A, Zarrindast MR. Scopolamine induced memory impairment; possible involvement of NMDA receptor mechanisms of dorsal hippocampus and/or septum. Behavioural Brain Research [Internet]. 2012 May 16 [cited 2024 May 2];231(1):1–10. Available from: https://pubmed.ncbi.nlm.nih.gov/22421366/
80. Flicker C, Serby M, Ferris SH. Scopolamine effects on memory, language, visuospatial praxis and psychomotor speed. Psychopharmacology. 1990;100(2):243–50.
81. Feiro O, Gould TJ. The interactive effects of nicotinic and muscarinic cholinergic receptor inhibition on fear conditioning in young and aged C57BL/6 mice. Pharmacology Biochemistry and Behavior. 2005 Feb 1;80(2):251–62.
82. Portero-Tresserra M, Del Olmo N, Martí-Nicolovius M, Guillazo-Blanch G, Vale-Martínez A. D-cycloserine prevents relational memory deficits and suppression of long-term potentiation induced by scopolamine in the hippocampus. European Neuropsychopharmacology. 2014 Nov 1;24(11): 1798–807.
83. Blokland A, Honig W, Raaijmakers WGM. Effects of intra-hippocampal scopolamine injections in a repeated spatial acquisition task in the rat. Psychopharmacology [Internet]. 1992 Nov [cited 2024 May 3];109(3):373–6. Available from: https://link.springer.com/article/10.1007/BF02245886
84. Chen KC, Baxter MG, Rodefer JS. Central blockade of muscarinic cholinergic receptors disrupts affective and attentional set-shifting. The European Journal of Neuroscience [Internet]. 2004 Aug [cited 2024 May 3];20(4):1081–8. Available from: https://pubmed.ncbi.nlm.nih.gov/15305877/
85. Singh P, Konar A, Kumar A, Srivas S, Thakur MK. Hippocampal chromatin-modifying enzymes are pivotal for scopolamine-induced synaptic plasticity gene expression changes and memory impairment. Journal of Neurochemistry [Internet]. 2015 Aug 1 [cited 2024 May 3];134(4):642–51. Available from: https://pubmed.ncbi.nlm.nih.gov/25982413/
86. Wu YY, Wang X, Tan L, Liu D, Liu XH, Wang Q, et al. Lithium attenuates scopolamine-induced memory deficits with inhibition of GSK-3β and preservation of postsynaptic components. Journal of Alzheimer's Disease. 2013 Jan 1;37(3):515–27.
87. SoukhakLari R, Moezi L, Pirsalami F, Ashjazadeh N, Moosavi M. Curcumin ameliorates scopolamine-induced mice memory retrieval deficit and restores hippocampal p-Akt and p-GSK-3β. European Journal of Pharmacology. 2018 Dec 15;841:28–32.
88. Olson ML, Cero IJ. Intrahippocampal Norleucine1-Angiotensin IV mitigates scopolamine-induced spatial working memory deficits. Peptides. 2010 Dec 1;31(12):2209–15.
89. Malik M, Rangel-Barajas C, Sumien N, Su C, Singh M, Chen Z, et al. The effects of sigma (σ1) receptor-selective ligands on muscarinic receptor antagonist-induced cognitive deficits in mice. British Journal of Pharmacology [Internet]. 2015 May 1 [cited 2024 May 2];172(10):2519–31. Available from: https://onlinelibrary.wiley.com/doi/full/10.1111/bph.13076
90. Madjid N, Tottie EE, Lüttgen M, Meister B, Sandin J, Kuzmin A, et al. 5-Hydroxytryptamine 1A receptor blockade facilitates aversive learning in mice: interactions with cholinergic and glutamatergic mechanisms. The Journal of Pharmacology and Experimental Therapeutics [Internet]. 2006 Feb [cited 2024 May 3];316(2):581–91. Available from: https://pubmed.ncbi.nlm.nih.gov/16223872/
91. Hirst WD, Stean TO, Rogers DC, Sunter D, Pugh P, Moss SF, et al. SB-399885 is a potent, selective 5-HT6 receptor antagonist with cognitive enhancing properties in aged rat water maze and novel object recognition models. European Journal of Pharmacology. 2006 Dec 28;553(1–3):109–19.

92. Brouillette J, Young D, During MJ, Quirion R. Hippocampal gene expression profiling reveals the possible involvement of Homerl and GABAB receptors in scopolamine-induced amnesia. Journal of Neurochemistry. 2007 Sep;102(6):1978–89.
93. Knox LT, Jing Y, Fleete MS, Collie ND, Zhang H, Liu P. Scopolamine impairs behavioural function and arginine metabolism in the rat dentate gyrus. Neuropharmacology. 2011 Dec 1;61(8):1452–62.
94. Hsieh MT, Hsieh CL, Lin LW, Wu CR, Huang GS. Differential gene expression of scopolamine-treated rat hippocampus-application of cDNA microarray technology. Life Sciences. 2003 Jul 11;73(8):1007–16.
95. Tang KS. The cellular and molecular processes associated with scopolamine-induced memory deficit: A model of Alzheimer's biomarkers. Life Sciences. 2019 Sep 15;233:116695.
96. Hritcu L, Stefan M, Brandsch R, Mihasan M. Enhanced behavioral response by decreasing brain oxidative stress to 6-hydroxy-l-nicotine in Alzheimer's disease rat model. Neuroscience Letters. 2015 Mar 30;591:41–7.
97. Liu W, Rabinovich A, Nash Y, Frenkel D, Wang Y, Youdim MBH, et al. Anti-inflammatory and protective effects of MT-031, a novel multitarget MAO-A and AChE/BuChE inhibitor in scopolamine mouse model and inflammatory cells. Neuropharmacology. 2017 Feb 1;113:445–56.
98. Chen BH, Park JH, Lee TK, Song M, Kim H, Lee JC, et al. Melatonin attenuates scopolamine-induced cognitive impairment via protecting against demyelination through BDNF-TrkB signaling in the mouse dentate gyrus. Chemico-Biological Interactions. 2018 Apr 1;285:8–13.
99. Hernández-Rodríguez M, Arciniega-Martínez IM, García-Marín ID, Correa-Basurto J, Rosales-Hernández MC. Chronic administration of scopolamine increased GSK3βP9, beta secretase, amyloid beta, and oxidative stress in the hippocampus of Wistar rats. Molecular Neurobiology [Internet]. 2020 Sep 1 [cited 2024 May 3];57(9):3979–88. Available from: https://pubmed.ncbi.nlm.nih.gov/32638218/
100. Kinney GG, Patino P, Mermet-Bouvier Y, Starrett JE, Gribkoff VK. Cognition-enhancing drugs increase stimulated hippocampal theta rhythm amplitude in the urethane-anesthetized rat. The Journal of Pharmacology and Experimental Therapeutics [Internet]. 1999 Oct 1 [cited 2024 May 2];291(1): 99–106. Available from: https://psycnet.apa.org/record/1999-11955-005
101. More SV, Kumar H, Cho DY, Yun YS, Choi DK. Toxin-induced experimental models of learning and memory impairment. International Journal of Molecular Sciences [Internet]. 2016 Sep 1 [cited 2024 May 3];17(9):1447. Available from: https://www.mdpi.com/1422-0067/17/9/1447/htm
102. Elsner M, Guldbakke B, Tiedge M, Munday R, Lenzen S. Relative importance of transport and alkylation for pancreatic beta-cell toxicity of streptozotocin. Diabetologia. 2000;43(12):1528–33.
103. Szkudelski T. The mechanism of alloxan and streptozotocin action in B cells of the rat pancreas. [cited 2024 Mar 26]; Available from: https://pubmed.ncbi.nlm.nih.gov/11829314/
104. Hoyer S, Muller D, Plaschke K. Desensitization of brain insulin receptor. Effect on glucose/energy and related metabolism. Journal of Neural Transmission, Supplement [Internet]. 1994 [cited 2024 Mar 25];(44):259–68. Available from: https://link.springer.com/chapter/10.1007/978-3-7091-9350-1_20
105. Stimulation of glycolytic key enzymes in cerebral cortex by … : NeuroReport [Internet]. [cited 2024 Mar 25]. Available from: https://journals.lww.com/neuroreport/abstract/1993/07000/Stimulation_of_glycolytic_key_enzymes_in_cerebral.39.aspx
106. Duelli R, Schröck H, Kuschinsky W, Hoyer S. Intracerebroventricular injection of streptozotocin induces discrete local changes in cerebral glucose utilization in rats. International Journal of Developmental Neuroscience [Internet]. 1994 Dec 1 [cited 2024 Mar 25];12(8):737–43. Available from: https://onlinelibrary.wiley.com/doi/full/10.1016/0736-5748%2894%2990053-1
107. Grieb P. Intracerebroventricular streptozotocin injections as a model of Alzheimer's disease: In search of a relevant mechanism. Molecular Neurobiology [Internet]. 2016 Apr 1 [cited 2024 Mar 25];53(3):1741. Available from:/pmc/articles/PMC4789228/
108. Qi CC, Chen XX, Gao XR, Xu JX, Liu S, Ge JF. Impaired learning and memory ability induced by a bilaterally hippocampal injection of streptozotocin in mice: Involved with the adaptive changes of synaptic plasticity. Frontiers in Aging Neuroscience [Internet]. 2021 Mar 1 [cited 2024 Mar 25];13:633495. Available from:/pmc/articles/PMC7957014/
109. Knezovic A, Osmanovic-Barilar J, Curlin M, Hof PR, Simic G, Riederer P, et al. Staging of cognitive deficits and neuropathological and ultrastructural changes in streptozotocin-induced rat model of Alzheimer's disease. Journal of Neural Transmission [Internet]. 2015 Apr 1 [cited 2024 Mar 25];122(4):577–92. Available from: https://link.springer.com/article/10.1007/s00702-015-1394-4

110. Salkovic-Petrisic M, Osmanovic-Barilar J, Brückner MK, Hoyer S, Arendt T, Riederer P. Cerebral amyloid angiopathy in streptozotocin rat model of sporadic Alzheimer's disease: A long-term follow up study. Journal of Neural Transmission [Internet]. 2011 May 30 [cited 2024 Mar 25];118(5):765–72. Available from: https://link.springer.com/article/10.1007/s00702-011-0651-4
111. Salkovic-Petrisic M, Knezovic A, Hoyer S, Riederer P. What have we learned from the streptozotocin-induced animal model of sporadic Alzheimer's disease, about the therapeutic strategies in Alzheimer's research. Journal of Neural Transmission (Vienna, Austria : 1996) [Internet]. 2013 Jan [cited 2024 Mar 25];120(1):233–52. Available from: https://pubmed.ncbi.nlm.nih.gov/22886150/
112. Sun P, Knezovic A, Parlak M, Cuber J, Karabeg M, Deckert J, et al. Long-term effects of intracerebroventricular streptozotocin treatment on adult neurogenesis in the rat hippocampus. Current Alzheimer Research [Internet]. 2015 Sep 16 [cited 2024 Mar 25];12(8):772–84. Available from: https://pubmed.ncbi.nlm.nih.gov/26159196/
113. Kamat PK, Kalani A, Rai S, Tota SK, Kumar A, Ahmad AS. Streptozotocin intracerebroventricular-induced neurotoxicity and brain insulin resistance: A therapeutic intervention for treatment of sporadic Alzheimer's disease (sAD)-like pathology. Molecular Neurobiology 2015 53:7 [Internet]. 2015 Aug 23 [cited 2024 Mar 25];53(7):4548–62. Available from: https://link.springer.com/article/10.1007/s12035-015-9384-y
114. Ravelli KG, Rosário B dos A, Camarini R, Hernandes MS, Britto LR. Intracerebroventricular streptozotocin as a model of Alzheimer's disease: Neurochemical and behavioral characterization in mice. Neurotoxicity Research [Internet]. 2017 Apr 1 [cited 2024 Mar 26];31(3):327–33. Available from: https://link.springer.com/article/10.1007/s12640-016-9684-7
115. Santos T de O, Mazucanti CHY, Xavier GF, da Silva Torrão A. Early and late neurodegeneration and memory disruption after intracerebroventricular streptozotocin. Physiology & Behavior. 2012 Oct 10; 107(3):401–13.
116. Bervian Bassani T, Bonato JM, Machado MMF, Cóppola-Segovia V, Moura ELR, Zanata SM, et al. Decrease in adult neurogenesis and neuroinflammation are involved in spatial memory impairment in the streptozotocin-induced model of sporadic Alzheimer's disease in rats. 2035.
117. Pinton S, da Rocha JT, Gai BM, Nogueira CW. Sporadic dementia of Alzheimer's type induced by streptozotocin promotes anxiogenic behavior in mice. Behavioural Brain Research. 2011 Sep 30;223(1):1–6.
118. C. Correia S, X. Santos R, S. Santos M, Casadesus G, C. LaManna J, Perry G, et al. Mitochondrial abnormalities in a streptozotocin-induced rat model of sporadic Alzheimer's disease. Current Alzheimer Research [Internet]. 2013 Apr 25 [cited 2024 Mar 26];10(4):406–19. Available from: https://pubmed.ncbi.nlm.nih.gov/23061885/
119. Paidi R, Nthenge-Ngumbau D, Singh R, Kankanala T, Mehta H, Mohanakumar K. Mitochondrial deficits accompany cognitive decline following single bilateral intracerebroventricular streptozotocin.
120. Poddar J, Singh S, Kumar P, Bali S, Gupta S, Chakrabarti S. Inhibition of complex I-III activity of brain mitochondria after intracerebroventricular administration of streptozotocin in rats is possibly related to loss of body weight. Heliyon. 2020 Jul 1;6(7):e04490.
121. Tristão Pereira C, Diao Y, Yin T, da Silva AR, Lanz B, Pierzchala K, et al. Synchronous nonmonotonic changes in functional connectivity and white matter integrity in a rat model of sporadic Alzheimer's disease. NeuroImage. 2021 Jan 15;225:117498.
122. Weinstock M, Shoham S. Rat models of dementia based on reductions in regional glucose metabolism, cerebral blood flow and cytochrome oxidase activity. Journal of Neural Transmission [Internet]. 2004 Mar [cited 2024 Mar 26];111(3):347–66. Available from: https://click.endnote.com/viewer?doi=10.1007%2Fs00702-003-0058-y&token=WzMxMjcwNiwiMTAuMTAwNy9zMDA3MDItMDAzLTAwNTgteSJd.g90eRnhw9cG4kOSXSPdHRECFY6o
123. Tong M, Leão R, Vimbela G V., Yalcin EB, Kay J, Krotow A, et al. Altered temporal lobe white matter lipid ion profiles in an experimental model of sporadic Alzheimer's disease. Molecular and Cellular Neurosciences [Internet]. 2017 Jul 1 [cited 2024 Mar 26];82:23–34. Available from: https://pubmed.ncbi.nlm.nih.gov/28438696/
124. Biswas J, Gupta S, Verma DK, Gupta P, Singh A, Tiwari S, et al. Involvement of glucose related energy crisis and endoplasmic reticulum stress: Insinuation of streptozotocin induced Alzheimer's like pathology. Cellular Signalling. 2018 Jan 1;42:211–26.
125. Yuliani T, Lobentanzer S, Klein J. Central cholinergic function and metabolic changes in streptozotocin-induced rat brain injury. Journal of Neurochemistry [Internet]. 2021 Sep 1 [cited 2024 Mar 26];158(6):1307–19. Available from: https://onlinelibrary.wiley.com/doi/full/10.1111/jnc.15155

126. Silva SSL, Tureck L V., Souza LC, Mello-Hortega J V., Piumbini AL, Teixeira MD, et al. Animal model of Alzheimer's disease induced by streptozotocin: New insights about cholinergic pathway. Brain Research. 2023 Jan 15;1799:148175.
127. Grieb P, Kryczka T, Fiedorowicz M, Frontczak-Baniewicz M, Walski M. Expansion of the Golgi apparatus in rat cerebral cortex following intracerebroventricular injections. Acta Neurobiologiae Experimentalis [Internet]. 2004 Dec 31 [cited 2024 Mar 25];64(4):481–9. Available from: https://ane.pl/index.php/ane/article/view/1531
128. Plaschke K, Müller D, Hoyer S. Insulin-resistant brain state (IRBS) changes membrane composition of fatty acids in temporal and entorhinal brain cortices of rats: relevance to sporadic Alzheimer's disease? Journal of Neural Transmission (Vienna, Austria : 1996) [Internet]. 2010 Dec [cited 2024 Mar 26];117(12):1419–22. Available from: https://pubmed.ncbi.nlm.nih.gov/21136123/
129. Sofic E, Salkovic-Petrisic M, Tahirovic I, Sapcanin A, Mandel S, Youdim M, et al. Brain catalase in the streptozotocin-rat model of sporadic Alzheimer's disease treated with the iron chelator–monoamine oxidase inhibitor, M30. Journal of Neural Transmission [Internet]. 2015 Apr 1 [cited 2024 Mar 26];122(4):559–64. Available from: https://link.springer.com/article/10.1007/s00702-014-1307-y
130. Carlos R, De Moraes M, Cristina G, Lima A, Aparecida C, Fernandes Cardinali E, et al. Benfotiamine protects against hypothalamic dysfunction in a STZ-induced model of neurodegeneration in rats. 2022 [cited 2024 Mar 26]; Available from: http://www.elsevier.com/open-access/userlicense/1.0/
131. Homolak J, Perhoc AB, Knezovic A, Barilar JO, Salkovic-Petrisic M. Failure of the brain glucagon-like peptide-1-mediated control of intestinal redox homeostasis in a rat model of sporadic Alzheimer's disease. Antioxidants [Internet]. 2021 Jul 1 [cited 2024 Mar 26];10(7). Available from:/pmc/articles/PMC8301063/
132. Homolak J, De Busscher J, Zambrano-Lucio M, Joja M, Virag D, Babic Perhoc A, et al. Altered secretion, constitution, and functional properties of the gastrointestinal mucus in a rat model of sporadic Alzheimer's disease. ACS Chemical Neuroscience [Internet]. 2023 Aug 2 [cited 2024 Mar 26];14(15):2667–82. Available from: https://pubs.acs.org/doi/full/10.1021/acschemneuro.3c00223
133. Song JZ, Cui SY, Cui XY, Hu X, Ma YN, Ding H, et al. Dysfunction of GABAergic neurons in the parafacial zone mediates sleep disturbances in a streptozotocin-induced rat model of sporadic Alzheimer's disease. Metabolic Brain Disease [Internet]. 2018 Feb 1 [cited 2024 Mar 26];33(1):127–37. Available from: https://link.springer.com/article/10.1007/s11011-017-0125-y
134. Humphrey CM, Hooker JW, Thapa M, Wilcox MJ, Ostrowski D, Ostrowski TD. Synaptic loss and gliosis in the nucleus Tractus Solitarii with streptozotocin-induced Alzheimer's disease. Brain research [Internet]. 2023 Feb 2 [cited 2024 Mar 26];1801:148202. Available from:/pmc/articles/PMC9840699/
135. Rostami F, Abbasi Z, Fereidoni M. Long-term follow-up of intracerebroventricular injection of streptozotocin-inducing pain sensitization. Behavioural Pharmacology [Internet]. 2022 Dec 1 [cited 2024 Mar 26];33(8):542–50. Available from: https://journals.lww.com/behaviouralpharm/fulltext/2022/12000/long_term_follow_up_of_intracerebroventricular.3.aspx
136. Bao J, Mahaman YAR, Liu R, Wang JZ, Zhang Z, Zhang B, et al. Corrigendum: Sex differences in the cognitive and hippocampal effects of streptozotocin in an animal model of sporadic AD. Frontiers in Aging Neuroscience [Internet]. 2019 Oct 15 [cited 2024 Mar 26];11:347. Available from:/pmc/articles/PMC6803784/
137. Gáspár A, Hutka B, Ernyey AJ, Tajti BT, Varga BT, Zádori ZS, et al. Corrigendum: Intracerebroventricularly injected streptozotocin exerts subtle effects on the cognitive performance of long-evans rats. Frontiers in Pharmacology [Internet]. 2023 Jan 5 [cited 2024 Mar 26];13. Available from: https://pubmed.ncbi.nlm.nih.gov/36686687/
138. Gáspár A, Hutka B, Judit Ernyey A, Tekla Tajti B, Tamás Varga B, Sándor Zádori Z, et al. Performance of the intracerebroventricularly injected streptozotocin Alzheimer's disease model in a translationally relevant, aged and experienced rat population. Scientific Reports | [Internet]. 123AD [cited 2024 Mar 26];12:20247. Available from: www.nature.com/scientificreports
139. Hsiao K, Chapman P, Nilsen S, Eckman C, Harigaya Y, Younkin S, et al. Correlative memory deficits, Abeta elevation, and amyloid plaques in transgenic mice. Science (New York, NY) [Internet]. 1996 [cited 2024 May 11];274(5284):99–102. Available from: https://pubmed.ncbi.nlm.nih.gov/8810256/
140. Heiland T, Zeitschel U, Puchades MA, Kuhn PH, Lichtenthaler SF, Bjaalie JG, et al. Defined astrocytic expression of human amyloid precursor protein in Tg2576 mouse brain. Glia [Internet]. 2019 Feb 1 [cited 2024 May 19];67(2):393–403. Available from: https://pubmed.ncbi.nlm.nih.gov/30485540/

141. Kawarabayashi T, Younkin LH, Saido TC, Shoji M, Ashe KH, Younkin SG. Age-dependent changes in brain, CSF, and plasma amyloid β protein in the Tg2576 transgenic mouse model of Alzheimer's disease. The Journal of Neuroscience [Internet]. 2001 Jan 1 [cited 2024 May 11];21(2):372. Available from:/pmc/articles/PMC6763819/
142. Duffy ÁM, Morales-Corraliza J, Bermudez-Hernandez KM, Schaner MJ, Magagna-Poveda A, Mathews PM, et al. Entorhinal cortical defects in Tg2576 mice are present as early as 2-4 months of age. Neurobiology of Aging [Internet]. 2015 Jan 1 [cited 2024 May 19];36(1):134–48. Available from: https://pubmed.ncbi.nlm.nih.gov/25109765/
143. Scopa C, Marrocco F, Latina V, Ruggeri F, Corvaglia V, La Regina F, et al. Impaired adult neurogenesis is an early event in Alzheimer's disease neurodegeneration, mediated by intracellular Aβ oligomers. Cell Death and Differentiation [Internet]. 2020 Mar 1 [cited 2024 May 19];27(3):934–48. Available from: https://pubmed.ncbi.nlm.nih.gov/31591472/
144. Hanafy AS, Lamprecht A, Dietrich D. Local perfusion of capillaries reveals disrupted beta-amyloid homeostasis at the blood-brain barrier in Tg2576 murine Alzheimer's model. Fluids and Barriers of the CNS [Internet]. 2023 Nov 22 [cited 2024 May 11];20(1):1. Available from: https://openurl.ebsco.com/contentitem/doi:10.1186%2Fs12987-023-00492-7?sid=ebsco:plink:crawler&id=ebsco:doi:10.1186%2Fs12987-023-00492-7
145. Spoleti E, Krashia P, La Barbera L, Nobili A, Lupascu CA, Giacalone E, et al. Early derailment of firing properties in CA1 pyramidal cells of the ventral hippocampus in an Alzheimer's disease mouse model. Experimental Neurology [Internet]. 2022 Apr 1 [cited 2024 May 19];350. Available from: https://pubmed.ncbi.nlm.nih.gov/34973962/
146. Alcantara-Gonzalez D, Chartampila E, Criscuolo C, Scharfman HE. Early changes in synaptic and intrinsic properties of dentate gyrus granule cells in a mouse model of Alzheimer's disease neuropathology and atypical effects of the cholinergic antagonist atropine. Neurobiology of Disease. 2021 May;152:105274.
147. Morello G, Guarnaccia M, La Cognata V, Latina V, Calissano P, Amadoro G, et al. Transcriptomic analysis in the hippocampus and retina of Tg2576 AD mice reveals defective mitochondrial oxidative phosphorylation and recovery by tau 12A12mAb treatment. Cells [Internet]. 2023 Sep 1 [cited 2024 May 11];12(18):2254. Available from: https://www.mdpi.com/2073-4409/12/18/2254/htm
148. Homolak J, Babic Perhoc A, Knezovic A, Osmanovic Barilar J, Virag D, Salkovic-Petrisic M. Exploratory study of gastrointestinal redox biomarkers in the presymptomatic and symptomatic Tg2576 mouse model of familial Alzheimer's disease: Phenotypic correlates and effects of chronic oral d-galactose. ACS Chemical Neuroscience [Internet]. 2023 Nov 15 [cited 2024 May 11];14(22):4013–25. Available from: https://pubmed.ncbi.nlm.nih.gov/37932005/
149. Kim JE, Roh YJ, Choi YJ, Lee SJ, Jin YJ, Song HJ, et al. Dysbiosis of fecal microbiota in Tg2576 mice for Alzheimer's disease during pathological constipation. International Journal of Molecular Sciences [Internet]. 2022 Dec 1 [cited 2024 May 11];23(23). Available from: https://pubmed.ncbi.nlm.nih.gov/36499254/
150. B. Szabo A, Cattaud V, Bezzina C, Dard RF, Sayegh F, Gauzin S, et al. Neuronal hyperexcitability in the Tg2576 mouse model of Alzheimer's disease – The influence of sleep and noradrenergic transmission. Neurobiology of Aging [Internet]. 2023 Mar 1 [cited 2024 May 11];123:35–48. Available from: https://pubmed.ncbi.nlm.nih.gov/36634385/
151. Torcinaro A, Ricci V, Strimpakos G, De Santa F, Middei S. Peripheral nerve impairment in a mouse model of Alzheimer's disease. Brain sciences [Internet]. 2021 Sep 1 [cited 2024 May 19];11(9). Available from: https://pubmed.ncbi.nlm.nih.gov/34573265/
152. Xia WF, Jung JU, Shun C, Xiong S, Xiong L, Shi XM, et al. Swedish mutant APP suppresses osteoblast differentiation and causes osteoporotic deficit, which are ameliorated by N-acetyl-L-cysteine. Journal of Bone and Mineral Research : The Official Journal of the American Society for Bone and Mineral Research. 2013 Oct;28(10):2122–35.
153. Pádua MS, Guil-Guerrero JL, Prates JAM, Lopes PA. Insights on the use of transgenic mice models in Alzheimer's disease research. International Journal of Molecular Sciences. 2024 Feb;25(5).
154. Duff K, Eckman C, Zehr C, Yu X, Prada CM, Perez-tur J, et al. Increased amyloid-beta42(43) in brains of mice expressing mutant presenilin 1. Nature. 1996 Oct;383(6602):710–3.
155. Matsuoka Y, Picciano M, Malester B, LaFrancois J, Zehr C, Daeschner JM, et al. Inflammatory responses to amyloidosis in a transgenic mouse model of Alzheimer's disease. The American Journal of Pathology. 2001 Apr;158(4):1345–54.
156. Takeuchi A, Irizarry MC, Duff K, Saido TC, Hsiao Ashe K, Hasegawa M, et al. Age-related amyloid beta deposition in transgenic mice overexpressing both Alzheimer mutant presenilin 1 and amyloid beta

precursor protein Swedish mutant is not associated with global neuronal loss. The American Journal of Pathology. 2000 Jul;157(1):331–9.
157. Arendash GW, King DL, Gordon MN, Morgan D, Hatcher JM, Hope CE, et al. Progressive, age-related behavioral impairments in transgenic mice carrying both mutant amyloid precursor protein and presenilin-1 transgenes. Brain Research. 2001 Feb;891(1–2):42 53.
158. Holcomb L, Gordon MN, McGowan E, Yu X, Benkovic S, Jantzen P, et al. Accelerated Alzheimer-type phenotype in transgenic mice carrying both mutant amyloid precursor protein and presenilin 1 transgenes. Nature Medicine. 1998 Jan;4(1):97–100.
159. Perez-Tur J, Froelich S, Prihar G, Crook R, Baker M, Duff K, et al. A mutation in Alzheimer's disease destroying a splice acceptor site in the presenilin-1 gene. Neuroreport. 1995 Dec;7(1):297–301.
160. Abd-Elrahman KS, Colson TLL, Sarasija S, Ferguson SSG. A M1 muscarinic acetylcholine receptor-specific positive allosteric modulator VU0486846 reduces neurogliosis in female Alzheimer's mice. Biomedicine & Pharmacotherapy = Biomedecine & Pharmacotherapie. 2024 Apr;173:116388.
161. Cai H, Wang Y, He J, Cai T, Wu J, Fang J, et al. Neuroprotective effects of bajijiasu against cognitive impairment induced by amyloid-β in APP/PS1 mice. Oncotarget. 2017 Nov;8(54):92621–34.
162. Oakley H, Cole SL, Logan S, Maus E, Shao P, Craft J, et al. Intraneuronal beta-amyloid aggregates, neurodegeneration, and neuron loss in transgenic mice with five familial Alzheimer's disease mutations: potential factors in amyloid plaque formation. The Journal of Neuroscience : The Official Journal of the Society for Neuroscience. 2006 Oct;26(40):10129–40.
163. Akhtar A, Gupta SM, Dwivedi S, Kumar D, Shaikh MF, Negi A. Preclinical models for Alzheimer's disease: Past, present, and future approaches. ACS Omega. 2022 Dec;7(51):47504–17.
164. Kimura R, Ohno M. Impairments in remote memory stabilization precede hippocampal synaptic and cognitive failures in 5XFAD Alzheimer mouse model. Neurobiology of Disease [Internet]. 2009;33(2): 229–35. Available from: https://www.sciencedirect.com/science/article/pii/S0969996108002581
165. Devi L, Ohno M. Genetic reductions of beta-site amyloid precursor protein-cleaving enzyme 1 and amyloid-beta ameliorate impairment of conditioned taste aversion memory in 5XFAD Alzheimer's disease model mice. The European Journal of Neuroscience. 2010 Jan;31(1):110–8.
166. Ohno M, Chang L, Tseng W, Oakley H, Citron M, Klein WL, et al. Temporal memory deficits in Alzheimer's mouse models: rescue by genetic deletion of BACE1. The European Journal of Neuroscience. 2006 Jan;23(1):251–60.
167. Girard SD, Baranger K, Gauthier C, Jacquet M, Bernard A, Escoffier G, et al. Evidence for early cognitive impairment related to frontal cortex in the 5XFAD mouse model of Alzheimer's disease. Journal of Alzheimer's Disease : JAD. 2013;33(3):781–96.
168. Sadleir KR, Kandalepas PC, Buggia-Prévot V, Nicholson DA, Thinakaran G, Vassar R. Presynaptic dystrophic neurites surrounding amyloid plaques are sites of microtubule disruption, BACE1 elevation, and increased Aβ generation in Alzheimer's disease. Acta Neuropathologica. 2016 Aug;132(2): 235–56.
169. Kalinin S, Polak PE, Lin SX, Sakharkar AJ, Pandey SC, Feinstein DL. The noradrenaline precursor L-DOPS reduces pathology in a mouse model of Alzheimer's disease. Neurobiology of Aging [Internet]. 2012;33(8):1651–63. Available from: https://www.sciencedirect.com/science/article/pii/S0197458011001540
170. Devi L, Ohno M. Phospho-eIF2α level is important for determining abilities of BACE1 reduction to rescue cholinergic neurodegeneration and memory defects in 5XFAD mice. PLoS One. 2010 Sep;5(9):e12974.
171. Eimer WA, Vassar R. Neuron loss in the 5XFAD mouse model of Alzheimer's disease correlates with intraneuronal Aβ42 accumulation and Caspase-3 activation. Molecular Neurodegeneration. 2013 Jan;8:2.
172. Jawhar S, Trawicka A, Jenneckens C, Bayer TA, Wirths O. Motor deficits, neuron loss, and reduced anxiety coinciding with axonal degeneration and intraneuronal Aβ aggregation in the 5XFAD mouse model of Alzheimer's disease. Neurobiology of Aging. 2012 Jan;33(1):196.e29-40.
173. Sethi M, Joshi SS, Webb RL, Beckett TL, Donohue KD, Murphy MP, et al. Increased fragmentation of sleep-wake cycles in the 5XFAD mouse model of Alzheimer's disease. Neuroscience. 2015 Apr;290:80–9.
174. Yoshiyama Y, Higuchi M, Zhang B, Huang SM, Iwata N, Saido TC, et al. Synapse loss and microglial activation precede tangles in a P301S tauopathy mouse model. Neuron. 2007 Feb;53(3):337–51.
175. Cao Q, Kumar M, Frazier A, Williams JB, Zhao S, Yan Z. Longitudinal characterization of behavioral, morphological and transcriptomic changes in a tauopathy mouse model. Aging. 2023 Nov;15(21): 11697–719.

176. Longo S, Messi ML, Wang ZM, Meeker W, Delbono O. Accelerated sarcopenia precedes learning and memory impairments in the P301S mouse model of tauopathies and Alzheimer's disease. Journal of Cachexia, Sarcopenia and Muscle. 2024 Aug;15(4):1358–75.
177. Arouche-Delaperche L, Cadoni S, Joffrois C, Labernede G, Valet M, César Q, et al. Dysfunction of the glutamatergic photoreceptor synapse in the P301S mouse model of tauopathy. Acta Neuropathologica Communications. 2023 Jan;11(1):5.
178. Martin SC, Joyce KK, Lord JS, Harper KM, Nikolova VD, Cohen TJ, et al. Sleep disruption precedes forebrain synaptic tau burden and contributes to cognitive decline in a sex-dependent manner in the P301S tau transgenic mouse model. eNeuro. 2024 Jun;11(6).
179. Sun Y, Guo Y, Feng X, Jia M, Ai N, Dong Y, et al. The behavioural and neuropathologic sexual dimorphism and absence of MIP-3α in tau P301S mouse model of Alzheimer's disease. Journal of Neuroinflammation. 2020 Feb;17(1):72.
180. Antón-Fernández A, Merchán-Rubira J, Avila J, Hernández F, DeFelipe J, Muñoz A. Phospho-tau accumulation and structural alterations of the Golgi apparatus of cortical pyramidal neurons in the P301S tauopathy mouse model. Journal of Alzheimer's Disease : JAD. 2017;60(2):651–61.
181. Alfaro-Ruiz R, Aguado C, Martín-Belmonte A, Moreno-Martínez AE, Merchán-Rubira J, Hernández F, et al. Alteration in the synaptic and extrasynaptic organization of AMPA receptors in the hippocampus of P301S tau transgenic mice. International Journal of Molecular Sciences. 2022 Nov;23(21).
182. Wang L, Liu M, Gao J, Smith AM, Fujioka H, Liang J, et al. Mitochondrial fusion suppresses tau pathology-induced neurodegeneration and cognitive decline. Journal of Alzheimer's Disease : JAD. 2021;84(3):1057–69.
183. Kim J, Choi IY, Duff KE, Lee P. Progressive pathological changes in neurochemical profile of the hippocampus and early changes in the olfactory bulbs of tau transgenic mice (rTg4510). Neurochemical Research. 2017 Jun;42(6):1649–60.
184. Kopeikina KJ, Polydoro M, Tai HC, Yaeger E, Carlson GA, Pitstick R, et al. Synaptic alterations in the rTg4510 mouse model of tauopathy. The Journal of comparative Neurology. 2013 Apr;521(6):1334–53.
185. Ishikawa A, Tokunaga M, Maeda J, Minamihisamatsu T, Shimojo M, Takuwa H, et al. In vivo visualization of tau accumulation, microglial activation, and brain atrophy in a mouse model of tauopathy rTg4510. Journal of Alzheimer's Disease : JAD. 2018;61(3):1037–52.
186. Yue M, Hanna A, Wilson J, Roder H, Janus C. Sex difference in pathology and memory decline in rTg4510 mouse model of tauopathy. Neurobiology of Aging. 2011 Apr;32(4):590–603.
187. Song L, Lu SX, Ouyang X, Melchor J, Lee J, Terracina G, et al. Analysis of tau post-translational modifications in rTg4510 mice, a model of tau pathology. Molecular Neurodegeneration. 2015 Mar;10:14.
188. Papanikolaou A, Rodrigues FR, Holeniewska J, Phillips KG, Saleem AB, Solomon SG. Plasticity in visual cortex is disrupted in a mouse model of tauopathy. Communications Biology. 2022 Jan;5(1):77.
189. Harrison IF, Ismail O, Machhada A, Colgan N, Ohene Y, Nahavandi P, et al. Impaired glymphatic function and clearance of tau in an Alzheimer's disease model. Brain : a Journal of Neurology. 2020 Aug;143(8):2576–93.
190. Keenan RJ, Daykin H, Chu J, Cornthwaite-Duncan L, Allocca G, Hoyer D, et al. Differential sleep/wake response and sex differences following acute suvorexant, MK-1064 and zolpidem administration in the rTg4510 mouse model of tauopathy. British Journal of Pharmacology. 2022 Jul;179(13):3403–17.
191. Oddo S, Caccamo A, Kitazawa M, Tseng BP, LaFerla FM. Amyloid deposition precedes tangle formation in a triple transgenic model of Alzheimer's disease. Neurobiology of Aging. 2003 Dec 1;24(8): 1063–70.
192. Janelsins MC, Mastrangelo MA, Oddo S, LaFerla FM, Federoff HJ, Bowers WJ. Early correlation of microglial activation with enhanced tumor necrosis factor-alpha and monocyte chemoattractant protein-1 expression specifically within the entorhinal cortex of triple transgenic Alzheimer's disease mice. Journal of Neuroinflammation [Internet]. 2005 Oct 18 [cited 2024 Apr 16];2(1):1–12. Available from: https://jneuroinflammation.biomedcentral.com/articles/10.1186/1742-2094-2-23
193. Kitazawa M, Oddo S, Yamasaki TR, Green KN, Laferla FM. Lipopolysaccharide-Induced Inflammation Exacerbates Tau Pathology by a Cyclin-Dependent Kinase 5-Mediated Pathway in a Transgenic Model of Alzheimer's Disease. 2005;
194. Olabarria M, Noristani HN, Verkhratsky A, Rodríguez JJ. Concomitant astroglial atrophy and astrogliosis in a triple transgenic animal model of Alzheimer's disease. Glia [Internet]. 2010 May 1 [cited 2024 Apr 20];58(7):831–8. Available from: https://onlinelibrary.wiley.com/doi/full/10.1002/glia.20967

195. Olabarria M, Noristani HN, Verkhratsky A, Rodríguez JJ. Age-dependent decrease in glutamine synthetase expression in the hippocampal astroglia of the triple transgenic Alzheimer's disease mouse model: mechanism for deficient glutamatergic transmission? 2011 [cited 2024 Apr 21]; Available from: http://www.molecularneurodegeneration.com/content/6/1/55
196. Oddo S, Caccamo A, Green KN, Liang K, Tran L, Chen Y, et al. Chronic nicotine administration exacerbates tau pathology in a transgenic model of Alzheimer's disease. Proceedings of the National Academy of Sciences of the United States of America [Internet]. 2005 Feb 2 [cited 2024 Apr 16];102(8): 3046. Available from:/pmc/articles/PMC549455/
197. Caccamo A, Oddo S, Billings LM, Green KN, Martinez-Coria H, Fisher A, et al. M1 receptors play a central role in modulating AD-like pathology in transgenic Mice. Neuron. 2006 Mar 2;49(5): 671–82.
198. Wander CM, Li YD, Bao H, Asrican B, Luo YJ, Sullivan HA, et al. Compensatory remodeling of a septo-hippocampal GABAergic network in the triple transgenic Alzheimer's mouse model. Journal of Translational Medicine [Internet]. 2023 Dec 1 [cited 2024 Apr 23];21(1):1–17. Available from: https://www.springermedicine.com/alzheimer-disease/alzheimer-disease/compensatory-remodeling-of-a-septo-hippocampal-gabaergic-network/25247378
199. Clinton LK, Billings LM, Green KN, Caccamo A, Ngo J, Oddo S, et al. Age-dependent sexual dimorphisms in cognition and stress response in the 3×Tg-AD mice. Neurobiology of Disease [Internet]. 2007 Oct [cited 2024 Apr 16];28(1):76. Available from:/pmc/articles/PMC2756084/
200. Hirata-Fukae C, Li HF, Hoe HS, Gray AJ, Minami SS, Hamada K, et al. Females exhibit more extensive amyloid, but not tau, pathology in an Alzheimer transgenic model. Brain Research. 2008 Jun 24; 1216:92–103.
201. Rodríguez JJ, Jones VC, Tabuchi M, Allan SM, Knight EM, LaFerla FM, et al. Impaired adult neurogenesis in the dentate gyrus of a triple transgenic mouse model of Alzheimer's disease. PLoS One [Internet]. 2008 Aug 13 [cited 2024 Apr 17];3(8). Available from:/pmc/articles/PMC2492828/
202. Liu Y, Bilen M, McNicoll MM, Harris RA, Fong BC, Iqbal MA, et al. Early postnatal defects in neurogenesis in the 3xTg mouse model of Alzheimer's disease. Cell Death & Disease [Internet]. 2023 Feb 1 [cited 2024 Apr 22];14(2). Available from:/pmc/articles/PMC9938901/
203. Yao J, Irwin RW, Zhao L, Nilsen J, Hamilton RT, Brinton RD. Mitochondrial bioenergetic deficit precedes Alzheimer's pathology in female mouse model of Alzheimer's disease. Proceedings of the National Academy of Sciences of the United States of America [Internet]. 2009 Aug 8 [cited 2024 Apr 17];106(34):14670. Available from:/pmc/articles/PMC2732886/
204. Fattoretti P, Balietti M, Casoli T, Giorgetti B, Di Stefano G, Bertoni-Freddari C, et al. Decreased numeric density of succinic dehydrogenase-positive mitochondria in CA1 pyramidal neurons of 3×Tg-AD mice. https://home.liebertpub.com/rej [Internet]. 2010 May 12 [cited 2024 Apr 20];13(2–3): 144–7. Available from: https://www.liebertpub.com/doi/10.1089/rej.2009.0937
205. de la Fuente-Muñoz CE, Rosas-Lemus M, Moreno-Castilla P, Bermúdez-Rattoni F, Uribe-Carvajal S, Arias C. Age-dependent decline in synaptic mitochondrial function is exacerbated in vulnerable brain regions of female 3×Tg-AD mice. International Journal of Molecular Sciences [Internet]. 2020 Nov 2 [cited 2024 Apr 22];21(22):1–13. Available from:/pmc/articles/PMC7699171/
206. Bertoni-Freddari C, Sensi SL, Giorgetti B, Balietti M, Di Stefano G, Canzoniero LMT, et al. Decreased presence of perforated synapses in a triple-transgenic mouse model of Alzheimer's disease. https://home.liebertpub.com/rej [Internet]. 2008 Apr 28 [cited 2024 Apr 17];11(2):309–13. Available from: https://www.liebertpub.com/doi/10.1089/rej.2008.0660
207. España J, Giménez-Llort L, Valero J, Miñano A, Rábano A, Rodriguez-Alvarez J, et al. Intraneuronal β-amyloid accumulation in the amygdala enhances fear and anxiety in Alzheimer's disease transgenic mice. Biological Psychiatry. 2010 Mar 15;67(6):513–21.
208. Desai MK, Sudol KL, Janelsins MC, Mastrangelo MA, Frazer ME, Bowers WJ. Triple-transgenic Alzheimer's disease mice exhibit region-specific abnormalities in brain myelination patterns prior to appearance of amyloid and tau pathology. Glia. 2009 Jan 1;57(1):54–65.
209. Cho SM, Lee S, Yang SH, Kim HY, Lee MJ, Kim HV, et al. Age-dependent inverse correlations in CSF and plasma amyloid-β(1-42) concentrations prior to amyloid plaque deposition in the brain of 3×Tg-AD mice OPEN. 2016 [cited 2024 Apr 23]; Available from: www.nature.com/scientificreports/
210. Bourasset F, Ouellet M, Tremblay C, Julien C, Do TM, Oddo S, et al. Reduction of the cerebrovascular volume in a transgenic mouse model of Alzheimer's disease. Neuropharmacology [Internet]. 2009 Mar [cited 2024 Apr 17];56(4):808–13. Available from: https://asu.elsevierpure.com/en/publications/reduction-of-the-cerebrovascular-volume-in-a-transgenic-mouse-mod
211. Lourenço CF, Ledo A, Barbosa RM, Laranjinha J. Neurovascular uncoupling in the triple transgenic model of Alzheimer's disease: Impaired cerebral blood flow response to neuronal-derived nitric oxide

signaling. Experimental Neurology [Internet]. 2017 Feb 1 [cited 2024 Apr 22];291:36–43. Available from: https://europepmc.org/article/med/28161255

212. Jullienne A, Quan R, Szu JI, Trinh M V., Behringer EJ, Obenaus A. Progressive vascular abnormalities in the aging 3×Tg-AD mouse model of Alzheimer's disease. Biomedicines [Internet]. 2022 Aug 1 [cited 2024 Apr 22];10(8). Available from:/pmc/articles/PMC9405684/
213. Wu M, Zhou F, Cao X, Yang J, Bai Y, Yan X, et al. Abnormal circadian locomotor rhythms and per gene expression in six-month-old triple transgenic mice model of Alzheimer's disease. Neuroscience Letters. 2018 May 29;676:13–8.
214. Sterniczuk R, Dyck RH, Laferla FM, Antle MC. Characterization of the 3×Tg-AD mouse model of Alzheimer's disease: Part 1. Circadian changes. Brain Research [Internet]. 2010 May 31 [cited 2024 Apr 20];1348:139–48. Available from: https://europepmc.org/article/med/20471965
215. Yin J, Tuo CM, Yu KY, Hu XH, Fan YY, Wu MN. Diurnal characteristics of the orexin system genes and its effects on pathology at early stage in 3×Tg-AD mice. Neuromolecular Medicine [Internet]. 2023 Oct 16 [cited 2024 Apr 22];25(4):632–43. Available from: https://europepmc.org/article/med/37843792
216. Gloria Y, Ceyzériat K, Tsartsalis S, Millet P, Tournier BB. Dopaminergic dysfunction in the 3×Tg-AD mice model of Alzheimer's disease. Scientific Reports [Internet]. 2021 Dec 1 [cited 2024 Apr 23];11(1). Available from:/pmc/articles/PMC8484608/

2 Rodent Models of Parkinson's Disease

Sahar Askari, Fatemeh Mokhtari Naseri, and Pegah Javadpour

2.1 INTRODUCTION

Parkinson's disease (PD) is one of the most prevalent neurodegenerative disorders, characterized by progressive motor dysfunction, non-motor symptoms, and the selective loss of dopaminergic neurons in the substantia nigra pars compacta. Despite extensive research, the precise mechanisms underlying PD remain incompletely understood, and effective disease-modifying therapies are yet to be developed. As such, preclinical research using animal models plays a crucial role in uncovering the pathophysiological processes of PD and testing potential therapeutic strategies. Rodents, particularly rats and mice, have become indispensable tools in PD research due to their well-characterized nervous systems, genetic manipulability, and cost-effectiveness. Various models have been developed to mimic different aspects of the disease, ranging from toxin-based approaches, such as 6-hydroxydopamine (6-OHDA) and rotenone models, to genetic models replicating familial PD mutations. Each model presents unique advantages and limitations, making it essential for researchers to select the most appropriate model based on their experimental objectives.

In the first section of this book, a brief definition of the pathological characteristics of PD will be provided. Subsequently, non-genetic and genetic models will be discussed in detail, highlighting their methodological considerations, strengths, and challenges.

2.2 PARKINSON'S DISEASE NEUROPATHOLOGY

PD is the second most common progressive neurodegenerative disorder, characterized by the loss of nigrostriatal dopamine neurons and the presence of α-synuclein inclusions (Lewy bodies). Unfortunately, PD is typically diagnosed only after more than 80% of nigrostriatal dopamine neurons have degenerated, coinciding with the onset of motor deficits (1). Resting tremor, rigidity, bradykinesia, akinesia, postural instability, and gait disturbances are well-known motor symptoms of PD. However, PD is more than just a dopamine-based motor disorder. Beyond dopaminergic dysfunction, patients often experience non-motor symptoms associated with disturbances in norepinephrine, serotonin, and non-aminergic systems (such as the cholinergic system). These symptoms include depression, anxiety, cognitive dysfunction, sleep disturbances, olfactory impairments, and gastrointestinal dysfunctions, many of which show little to no response to dopaminergic therapies. Importantly, such non-motor symptoms are now recognized as early indicators of PD and may appear several years, or even decades, before the onset of motor impairments (1). Therefore, to better capture the expanding definition of PD, various animal models are needed to address the diverse aspects of this heterogeneous disorder beyond motor deficits.

2.3 RODENT MODELS OF PARKINSON'S DISEASE

2.3.1 Non-Genetic Models

2.3.1.1 Reserpine

Reserpine is a well-known inhibitor of vesicular monoamine transporter-2 (VMAT2), a protein responsible for storing monoamines such as noradrenaline, serotonin, and dopamine. By inhibiting

DOI: 10.1201/9781032676524-3

VMAT2, reserpine depletes these monoamines, leading to functional alterations (2). The administration of reserpine to rodents can induce short-term PD-like symptoms, which can be alleviated by L-DOPA (levodopa) treatment (3).

Although this model effectively mimics some aspects of PD, it has been criticized for its lack of dopamine selectivity and the absence of dopamine neuron degeneration (4,5). The depletion of other monoamines, such as noradrenaline and serotonin, also contributes to the observed symptoms, reflecting some non-dopaminergic aspects of PD. Findings in mice and rats confirm that reserpine administration induces akinesia and hind limb rigidity, with these effects persisting for 24 hours in reserpine-treated rats (6). At doses of 0.5 and 1 mg/kg, the subcutaneous administration of reserpine for four consecutive days resulted in a notable increase in vacuous chewing movements and a higher frequency of akinesia. These behavioral alterations were associated with a reduction in tyrosine hydroxylase (TH) immunoreactivity in the striatum, which was evident on days 2 and 20 following the final administration of 1 mg/kg (7). Additionally, experimental studies have demonstrated a significant decline in locomotor activity in both rats and mice subjected to reserpine treatment at a dose of 0.6 mg/kg via subcutaneous injection (0.6 mg/kg, subcutaneous) (8).

2.3.1.2 Haloperidol

Haloperidol, a widely used antipsychotic drug, exerts its pharmacological effects by inhibiting dopamine transmission in the striatum through the blockade of dopamine D2 and D1 receptors. Additionally, haloperidol exerts inhibitory effects on adenosine receptors, particularly A2A receptors, which are known to modulate striatal dopamine signaling.

Although haloperidol, like reserpine, does not induce the formation of Lewy bodies, it remains a commonly employed model for evaluating symptom-related effects of non-dopaminergic agents, such as adenosine A2A/A1 antagonists, mGlu4-positive allosteric modulators, and mGlu7 agonists (9). Haloperidol administration can lead to the development of short-term Parkinsonian-like symptoms in rodents and rabbits, which are alleviated by L-DOPA treatment (3). Various motor deficits have been observed in this model, including catalepsy (10), akinesia, rigidity, and hypo-locomotion (11–13). These symptoms reflect the multifaceted effects of haloperidol as both a dopaminergic and adenosine receptor inhibitor.

2.3.1.3 6-Hydroxydopamine (6-OHDA)

6-OHDA, a structural homolog of dopamine and noradrenaline, exhibits a toxic impact on catecholaminergic neurons. This characteristic makes it a valuable tool for researchers seeking to induce PD in animal models.

The first application of 6-OHDA dates back to 1959 when Porter et al. demonstrated its efficacy in causing efficient and enduring depletion of noradrenaline in sympathetic nerves of the heart. Subsequently, in 1968, Ungerstedt administered bilateral injections of 6-OHDA into the substantia nigra pars compacta (SNpc), revealing anterograde degeneration of dopaminergic neurons (14). This marked the initiation of 6-OHDA's utilization in preclinical studies for PD.

The neurotoxin 6-OHDA permeates dopaminergic and noradrenergic neurons via monoamine transporters. Once internalized, it accumulates in mitochondria, impairing the function of complex I. Consequently, this disruption triggers an augmented production of free radicals, including superoxide, hydrogen peroxide, and hydroxyl radicals, alongside a reduction in levels of antioxidants such as glutathione and superoxide dismutase. Consequently, this disturbance in redox homeostasis precipitates lipid peroxidation, oxidative damage to DNA and proteins, and the release of cytochrome c. Additionally, the neurotoxin hampers complex IV of mitochondria, leading to a decline in mitochondrial membrane potential. These cumulative effects contribute to mitochondrial dysfunction, neuronal death, and the activation of astrocytes and microglia (6).

The catalyst for the toxic reactions of 6-OHDA in neurons is iron. As a result, the use of iron-chelating agents, including monoamine oxidase (MAO)-B inhibitors and vitamin E, proves instrumental in preventing induced damage. Furthermore, various antioxidant agents have demonstrated a protective effect against such damage (6).

It is crucial to note that 6-OHDA infiltrates both dopaminergic and noradrenergic neurons. To selectively target dopaminergic neurons, it is recommended to administer a noradrenergic reuptake inhibitor, such as desipramine, 30–60 minutes before injection of 6-OHDA (15).

One significant drawback of this toxin is its inability to cross the blood–brain barrier (BBB). Consequently, specialized skills and surgical instruments are requisite for direct injection into the brain (16).

To keep solutions with 6-OHDA safe, it's important to shield them from light and add ascorbic acid. This helps prevent the 6-OHDA from breaking down, as it's sensitive to light and oxidation.

However, recent demonstrations indicate that ascorbic acid not only acts as an antioxidant but also enhances the selectivity and toxicity of 6-OHDA.

6-OHDA can cause a bilateral lesion when injected into both hemispheres or the ventricles, but a unilateral lesion is more commonly used. Different profiles of cell death can occur depending on the injection site and dose.

In accordance with the study's rationale, 6-OHDA can be administered unilaterally to the SNpc, medial forebrain bundle (MFB), or striatum. Unilateral injection results in asymmetric motor deficits and rotation behavior. On the other hand, bilateral injection induces severe manifestations such as aphagia, adipsia, seizures, and high mortality (16,17).

The injection of the toxin into the striatum precipitates a gradual retrograde degeneration of nigrostriatal neurons spanning a duration of one to three weeks. Axonal denervation becomes apparent within 3 hours post-injection in the striatum. However, the reduction of TH-positive neurons in the SNpc is seen several days after surgery (16,17).

In contrast, the injection of 6-OHDA into the MFB or SNpc (called generally the MFB model) induces a severe and rapid neurodegenerative response. In this model, striatal dopamine levels decrease significantly after three to four days of the lesion. Studies show that the remaining dopamine content is less than 20% of normal. When 6-OHDA is injected into the MFB, dopamine content in the striatum is completely lost within three weeks. The number of dopaminergic neurons in the SNpc gradually decreases, with nearly complete loss (90%) after about five weeks. This suggests that neurodegeneration is still occurring even when the striatum is fully deprived of dopamine. Injections into these regions result in a spectrum of motor and non-motor symptoms, including gastrointestinal dysfunction and cognitive impairment (16–18).

The MFB model and the striatal model serve distinct purposes in the study of PD. The striatal model is more pertinent for gaining insights into the specific mechanisms underlying cell death in PD. In contrast, the MFB model is better suited for investigating the outcomes of dopaminergic neuronal death and assessing therapeutic approaches aimed at alleviating motor symptoms associated with PD (17).

In postmortem Parkinson's brains, dopaminergic neurons show signs of both apoptotic and non-apoptotic cell death.

The molecular pathways leading to cell death induced by 6-OHDA have primarily been investigated in cell cultures, with limited exploration in vivo.

Oxidative stress triggered by 6-OHDA in PC12 cells seems to occur via an increase in SAPK/JNK phosphorylation and a decrease in PI3K phosphorylation.

Reduced phosphorylation of PI3K leads to the activation of the pro-apoptotic protein family, including Bax, and inhibits the anti-apoptotic protein Bcl-2. This, in turn, results in decreased survivin levels and increased release of Cytochrome c from mitochondria, subsequently triggering the activation of caspase 9 and caspase 3. Activation of caspase 3 enhances the cleavage of PARP, ultimately culminating in apoptosis(19).

Moreover, 6-OHDA-containing ascorbic acid induces cell death by disrupting intracellular calcium homeostasis, leading to calpain activation and subsequent mitochondrial damage (20).

Additionally, studies have demonstrated that both p53 and the BH3-only protein, p53-upregulated mediator of apoptosis (PUMA), play crucial roles in 6-OHDA-induced cell death (21).

Furthermore, there is a suggested disruption of STAT3 signaling by 6-OHDA, which is believed to contribute to neuronal death in PD (22).

Ferroptosis, another form of cell death, appears to be induced by 6-OHDA. As indicated, iron plays a significant role in the toxicity of 6-OHDA. It is well-established that ferroptosis, a type of neuronal cell death, is dependent on iron and has been observed in recent studies on PD. Research conducted on PC12 cells exposed to 6-OHDA has verified the accumulation of intracellular Fe^{2+}. Apart from the increase in Fe^{2+}, other signs of ferroptosis include changes in expression of the GPX4, lipid peroxidation, glutathione levels, and alterations in mitochondrial membrane potential. Significant alterations in all these factors have been reported, confirming the activation of ferroptosis following exposure to 6-OHDA (23).

Microglial activation contributes to the degeneration of dopaminergic neurons, contrasting with the beneficial effects associated with astroglial activation. In this context, the exogenous administration of fractalkine (CX3CL1), a neuron-derived suppressor of microglial activation, has been shown to protect dopaminergic neurons against 6-OHDA-induced toxicity. This underscores the preliminary role of microglia in the toxicity associated with 6-OHDA (17).

6-OHDA disrupts not only mitochondrial transport but also the movement of vesicles, including those labeled with synaptophysin. The temporal sequence of events indicates that microtubule disruption and autophagy occur after transport dysfunction but precede dopamine cell death following 6-OHDA treatment. It seems that early transport dysfunction mediated by reactive oxygen species (ROS) significantly contributes to the initiation of axonal degeneration in response to 6-OHDA treatment (24).

The administration of 6-OHDA appears to induce an elevation in glutamate and δ-aminobutyric acid (GABA), accompanied by a reduction in glutamine levels within the striatum. This implies a probable alteration in the steady-state equilibrium of the glutamine-glutamate (Gln-Glu) cycle between astrocytes and neurons. Such a modification in the Gln-Glu cycle has the potential to disturb the delicate balance between excitatory and inhibitory processes in the brain, possibly resulting in enduring abnormalities in glutamatergic and GABAergic activities (17).

The 6-OHDA model is distinguished by the distinctive rotational behavior observed in affected rodents following the administration of dopaminomimetics. This behavioral response is directly linked to the extent of the nigrostriatal lesion. It should be noted that it does not precisely reflect the extent of dopaminergic cell loss (17).

In response to reduced dopamine levels, various compensatory mechanisms, including an increase in dopamine D2 postsynaptic receptor density, are triggered in response to the lesion.

This heightened sensitivity to dopamine on the lesioned side results in an asymmetry between the two striata, leading to an uneven response to systemically administered dopaminergic drugs. For instance, administration of dopaminergic agonists like apomorphine or L-DOPA induces contralateral rotation, whereas drugs that stimulate dopamine release, such as amphetamines, prompt the rodent to turn toward the direction of the lesion. This discrepancy arises because the non-lesioned side can release more dopamine than the lesioned side (17).

Rotational behavior can be quantified to assess the anti-Parkinsonian effects of novel drugs, evaluate the effectiveness of transplantation and gene therapies, and establish a model for L-DOPA motor fluctuations following prolonged treatment (17).

It is important to highlight that 6-OHDA has been discovered in both the human caudate nucleus (25) and the urine of PD patients undergoing L-DOPA treatment (26). This suggests a potential involvement of 6-OHDA in PD pathogenesis as a natural metabolite of dopamine.

Indirect evidence also indicates heightened oxidative stress in the dopaminergic neurons of PD patients, with reported decreases in glutathione peroxidase and catalase activity, reduced glutathione levels, and evidence of lipid peroxidation and DNA oxidative damage. As an inducer of oxidative stress, 6-OHDA may replicate some features observed in the course of PD.

These animals exhibit motor symptoms such as bradykinesia and a diminished capacity to execute motor tasks on the side of the body affected. It's noteworthy that older animals appear to be more susceptible to the effects of 6-OHDA compared to their younger counterparts.

Motor behavioral dysfunction induced by 6-OHDA can be reversed with levodopa treatment. Prolonged administration of levodopa in animals with 6-OHDA lesions leads to involuntary movements like levodopa-induced dyskinesia, a significant side effect observed in PD patients undergoing levodopa therapy. This model proves valuable for translational studies focused on dopamine replacement and therapies targeting levodopa-induced dyskinesia.

Many research labs have noted that the intentional reduction of dopaminergic neurons using 6-OHDA, in addition to causing motor symptoms, results in deficits in various functions such as cognition, gut motility, enteric nervous system (ENS) function, olfaction, and the sleep-wake cycle. These observations suggest that the decline in dopamine levels may play a role, at least partially, in the development of non-motor symptoms in PD (27).

It's essential to highlight that 6-OHDA does not generate Lewy body-like inclusions in the nigrostriatal pathway.

The chemical structure of 6-OHDA, its injection sites, and the resulting effects in rodents are presented in Table 2.1.

2.3.1.4 1-Methyl-4-phenyl-1,2,3,6-tetrahydropyridine (MPTP)

1-Methyl-4-phenyl-1,2,3,6-tetrahydropyridine (MPTP) is a byproduct of the synthesis of 1-methyl-4-phenyl-propionoxypiperidine (MPPP). Its discovery in 1892 was accidental, following the identification of patients exhibiting neurological symptoms resembling idiopathic PD. Remarkably, these patients responded dramatically to levodopa, which sparked further investigation. The common link between these patients was their intravenous self-administration of synthetic heroin (MPPP) (28–30).

MPTP is a neurotoxic contaminant that damages dopaminergic neurons in a dose-dependent manner within the substantia nigra and putamen (31). Due to its high lipophilicity, MPTP readily crosses the BBB, where it is first taken up by glial cells. Inside these cells, MPTP is converted to 1-methyl-4-phenyl-2,3-dihydropyridinium ($MPDP^+$), which is ultimately metabolized into the active neurotoxin 1-methyl-4-phenylpyridinium (MPP^+) (4).

MPP^+ enters dopaminergic neurons selectively via the dopamine transporter, accumulates, and is packed into synaptic vesicles containing dopamine through the vesicular monoamine transporter (32). Once inside the neurons, MPP+ disrupts the mitochondrial electron transport chain, reducing

TABLE 2.1
Key Characteristics of 6-Hydroxydopamine Exposure in Rodents

Aspect	Details
Chemical Structure	Structural analog of dopamine and noradrenaline.
Injection Sites	Substantia nigra pars compacta, Medial Forebrain Bundle, Striatum
Effects	• Cognitive impairments • Disturbances in sleep-wake cycles • Gut motility changes • Iron accumulation • Mitochondrial dysfunction • Motor deficits • Neuroinflammation • Olfactory impairment • Oxidative stress • Neuronal cell death

ATP production, enhancing oxidative stress, and ultimately leading to neuronal death. Interestingly, mice lacking dopamine transporters are protected from MPTP-induced toxicity (33).

In MPTP-treated mice, α-synuclein aggregation is observed, mirroring the pathology seen in PD (34). Furthermore, motor deficits such as impaired forepaw stride length, which correlate with dopamine depletion in the basal ganglia, are apparent. These motor deficits improve following acute levodopa administration (35). Notably, some extra-nigral pathology is present, as MPTP-treated mice show prolonged latency in finding a buried pellet, potentially due to dopamine depletion in the olfactory tubercle (36). Sleep disturbances are also common, with an increase in paradoxical sleep episodes and bouts 20 days post-exposure (37,38). The ENS is also impacted by MPTP, leading to a reduction in dopaminergic neurons, which in turn impairs the relaxation of proximal colon muscles. Interestingly, despite this impairment, colon motility has been reported to increase, while gastric emptying and small intestine transit remain unaffected (39). In terms of anxiety-like behaviors, no significant differences have been observed between MPTP-treated and saline-treated animals in standard tests, such as light-dark exploration, tail suspension, or sucrose preference (40).

Interestingly, rodents are generally less sensitive to MPTP toxicity compared to primates, exhibiting greater resistance to MPTP-induced neurodegeneration (41). Among rodents, mice, especially the C57BL/6 strain, are more sensitive to MPTP administration than rats, which are relatively resistant and show minimal Parkinsonian features unless treated with significantly higher doses (42). This difference in susceptibility is partly attributed to the higher levels of MAO in rat brains, an enzyme responsible for converting MPTP to MPP+. The lower MAO levels in mice facilitate greater brain penetration of MPTP, making the mouse model more commonly used in PD research (43,44).

Due to logistical, ethical, and economic constraints, primate research on MPTP has become increasingly limited. As a result, the MPTP mouse model has become the preferred model for investigating PD (45). MPTP can be administered through various routes in rodents, including oral gavage, stereotaxic injection, and systemic methods such as subcutaneous, intramuscular, intravenous, and intraperitoneal administration (46). Among different routes, the most common route in mice is the intraperitoneal injection, which primarily targets dopaminergic neurons in the SNpc (31). It should be noted that motor deficits in MPTP-treated mice are transient and typically recover within a few days post-acute injection, which imposes limitations on the duration of behavioral experiments (47). The administration schedule also affects the progression of dopaminergic degeneration. The acute model typically involves a moderate dose (20 mg/kg) of MPTP injected four times at 2-hour intervals, resulting in a rapid loss of dopaminergic neurons, with 90% striatal dopamine depletion and approximately 70% loss of SNpc dopaminergic neurons. However, this model does not produce α-synuclein aggregation or apoptotic cell death, with the symptoms stabilizing within a week (48). In contrast, in the sub-acute model, which is also called "delayed degeneration", daily injection of a 30 mg/kg MPTP for five consecutive days results in 40–50% striatal dopamine depletion and 30–40% cell loss (49). Additionally, a more progressive MPTP model has been developed using continuous minipump infusions of low MPTP doses. This approach results in the degeneration of both dopaminergic and noradrenergic neurons, depletion of dopamine and norepinephrine, and the formation of neuronal inclusions resembling lewy bodies in the substantia nigra and locus coeruleus (34). Furthermore, studies in rodent models have demonstrated that intranasal administration of MPTP induces cognitive impairment (50,51). Notably, the neurotoxic effects of MPTP, including dopaminergic neuronal death, neuroinflammation, truncal dystonia, catalepsy, and hindlimb clasping behavior, have been linked to the activation of the microglial NLRP3 inflammasome. Importantly, these pathological features were mitigated in models with NLRP3 deficiency, highlighting the inflammasome's potential role in MPTP-induced neurodegeneration (52).

For the last tip, when selecting the appropriate mouse strain for MPTP studies, factors such as gender, body weight, and age must be carefully considered, as these parameters can influence the outcomes of the model (53,54). For example, male mice weighing at least 22 g and at least eight weeks old are recommended for MPTP studies in the mouse model (48).

TABLE 2.2
Key Characteristics of MPTP Exposure in Rodents

Aspect	Details
Origin	Byproduct of 1-methyl-4-phenyl-propionoxypiperidine (MPPP) synthesis
Administration Routes	Oral (Gavage), Stereotaxic, Subcutaneous, Intramuscular, Intravenous, Intraperitoneal
Effects	• Disturbances in sleep-wake cycles • Enteric nervous system dysfunction • Mitochondrial dysfunction • Motor deficits • Olfactory impairment • Oxidative stress

The summarized key features of MPTP administration in rodents, its origin, and its exposure methods are presented in Table 2.2.

2.3.1.5 Rotenone

Rotenone is a naturally occurring pesticide and piscicide derived from Leguminosae plants. As a potent inhibitor of mitochondrial complex I, it has been widely used in experimental models of PD due to its neurotoxic effects. The chronic rotenone model provides critical insights into the progressive nature of PD by replicating its slow evolution and prolonged prodromal phase.

Daily intraperitoneal administration of a low dose (2.75 mg/kg) to adult male rats for 21 days (weekdays only) induces a time-dependent progression of behavioral and neurochemical changes. Motor impairments, including bradykinesia in the open field test and coordination deficits, become evident after 14 days of exposure. These deficits correlate with nigrostriatal dopaminergic degeneration and increased α-synuclein accumulation, both hallmark features of PD. Notably, several alterations precede the onset of motor symptoms, including gastrointestinal dysfunction, changes in whole-body composition, and olfactory and taste impairments (55,56). These findings emphasize the progressive nature of PD pathology in the rotenone model and its relevance for studying both early and late-stage disease mechanisms (57).

A timeline study using subcutaneous administration of rotenone (4 mg/kg daily for 35 days) in C57BL/6 mice identified three pathological stages: cognitive impairments by week 2, subtle motor deficits in weeks 3–4, and pronounced motor dysfunction from week 4 onward. These findings further support the staged progression of PD-like symptoms in rotenone-treated animals (58).

Beyond its effects on the nigrostriatal dopaminergic system, rotenone also adversely impacts hippocampal and cortical neurons, leading to significant cognitive impairments in mice. These deficits mirror the non-motor symptoms observed in PD and suggest a broader neurotoxic profile beyond the basal ganglia. Recent studies indicate that microglia-mediated neuroinflammation is a key contributor to rotenone-induced cognitive dysfunction (56,59). Moreover, recognition memory decline has been linked to significant alterations in AMPA receptor subunits, including upregulation of GluR1 and downregulation of GluR2 in the hippocampus (60).

Functional abnormalities in catecholaminergic neurons have also been reported following subcutaneous administration of rotenone (2 mg/kg/day via a minipump for 10 days), replicating the pattern observed in the putamen of PD patients (61). Moreover, rotenone has been shown to disrupt the BBB in a dose-dependent manner, leading to increased Evans blue extravasation, fibrinogen accumulation, and a reduction in tight junction protein expression in the hippocampus and cortex (62). Beyond its neurotoxic effects, rotenone induces metabolic disturbances linked to obesity and hyperinsulinemia, suggesting systemic consequences beyond the central nervous system (63). Additionally, rotenone administration disrupts gut microbiota composition and intestinal barrier integrity, contributing to systemic inflammation and neurodegeneration. One key mechanism involved in this process is the activation of the CCAAT/enhancer binding protein β/asparagine endopeptidase (C/EBPβ/AEP) pathway, stimulated by bacterial endotoxins. Notably, C/EBPβ/AEP signaling enhances α-synuclein transcription, linking peripheral gut dysfunction to central pathological features of PD (64).

According to studies, aging plays a critical role in influencing susceptibility to α-synuclein pathology, neuroinflammation, and neurodegeneration in rotenone-induced PD models. A study using intranasal rotenone microemulsion in 12-month-old female BALB/c mice over six weeks demonstrated a progressive onset of non-motor and motor deficits, reinforcing the role of age in enhancing rotenone-induced neuropathology. Olfactory dysfunction emerged as an early symptom after four weeks of exposure, while motor impairments became apparent only after six weeks, highlighting a time-dependent progression of pathology. Postmortem analyses confirmed widespread α-synuclein accumulation in the olfactory bulb, striatum, substantia nigra, and cortex, along with significant glial activation and neurodegeneration in these regions. Notably, these neurotoxic effects were absent in younger (three-month-old) mice, even after a prolonged 9-week exposure, underscoring the heightened vulnerability of aged brains to rotenone-induced neurotoxicity. Additionally, iron accumulation in the substantia nigra of aged rotenone-exposed mice suggests a potential link between disrupted iron homeostasis and oxidative stress-driven neurodegeneration (65).

Rotenone is administered via various routes, each with distinct advantages and limitations. Some studies utilize stereotaxic surgery for direct brain administration. Bilateral stereotaxic injection of rotenone (3 μg) into the MFB of rats results in significant dopaminergic neuronal loss and behavioral impairments, including decreased locomotion and increased catalepsy (66). Additionally, unilateral infusion of a low rotenone dose (2 μg) into the ventral tegmental area and substantia nigra reduces dopamine levels and induces α-synuclein expression on the injected side (67). In addition, rotenone is administered both systematically and intranasally. Due to its high lipophilicity, rotenone readily crosses the BBB, reaching peak concentrations in the CNS within 15 minutes following a single intravenous injection (68).

So, rotenone-based models effectively replicate the progressive neurodegeneration seen in PD, including both motor and non-motor symptoms.

The effects of rotenone administration in rodents, along with its mechanism of action and exposure methods, are summarized in Table 2.3.

TABLE 2.3
Key Characteristics of Rotenone Exposure in Rodents

Aspect	Details
Mechanism of Action	Inhibits mitochondrial complex I, leading to oxidative stress and neurotoxicity
Mode of Administration	Intraperitoneal, Subcutaneous, Intranasal, Intravenous, Intracerebral
Effects	• α-Synuclein accumulation • Blood–brain barrier disruption • Cognitive deficits • Dopaminergic degeneration • Gastrointestinal dysfunction • Microglia activation • Motor impairments • Olfactory and taste dysfunction

2.3.2 Genetic Models

Recent research suggests that approximately 10% of PD cases have a familial origin. Several gene mutations associated with familial PD have been identified, including α-synuclein, Parkin-1, PTEN-induced putative kinase 1 (PINK1), DJ-1, and LRRK2. These genetic variations have been instrumental in the development of transgenic rodent models, enabling researchers to investigate the molecular mechanisms underlying familial PD.

2.3.2.1 α-Synuclein-Based Genetic Model

The aggregation of α-synuclein is a hallmark of PD, contributing to Lewy body formation and disrupting the function of mitochondria, the Golgi apparatus, and the endoplasmic reticulum, ultimately leading to nigrostriatal dopaminergic degeneration (69). The SNCA gene, which encodes α-synuclein, was the first identified gene linked to familial PD (70). To date, eight missense mutations in SNCA (A53T/E/V, A30P/G, E46K, H50Q, and G51D) have been associated with α-synuclein accumulation and aggregation in the brain (71). Accordingly, several transgenic models, including those carrying E46K, A30P, and A53T mutations, have been developed to study α-synuclein pathology in PD. The E46K human α-synuclein transgenic mouse model exhibits elevated α-synuclein expression and aggregation in multiple brain regions, including the spinal cord, brain stem, deep cerebellar nuclei, motor cortex, and most of the thalamus. These pathological changes are accompanied by significant motor deficits, which typically manifest between 15 and 19 months of age (72). Furthermore, the E46K α-synuclein transgenic rat model demonstrates reduced dopamine metabolites in the striatum, increased α-synuclein accumulation and aggregation, and heightened oxidative stress in nigral dopamine neurons. However, despite these neurochemical and pathological alterations, no significant motor impairments have been reported in this model (73). Another α-synuclein-based model involves A30P mutant α-synuclein transgenic mice, which have been generated using different approaches. These include the targeted A30P point mutation in the wild-type gene (Sncatm (A30P)), or overexpression of the human A30P α-synuclein gene under either the prion promoter (Prnp-A30P) (74) or the Ty1 promoter (Ty1-A30P) (75). An accumulation of α-synuclein was observed in the brains of all these mice as they aged. Motor disabilities and abnormalities associated with early PD symptoms were reported in Prnp-A30P and Ty1-A30P transgenic mice (74,75). However, motor deficits were not observed in Sncatm (A30P) mice (74). The A53T mutation in α-synuclein is one of the most widely studied genetic models of PD, exhibiting mid-to-late onset of symptoms. The M83 transgenic mouse model, the first to overexpress human A53T-mutated α-synuclein, displays α-synuclein aggregation and inclusion formation in various brain regions (76,77). The highest density of α-synuclein inclusions has been observed in the spinal cord, brain stem, deep cerebellar nuclei, white matter, and specific thalamic regions (78). However, despite these pathological changes, M83 homozygous mice do not exhibit significant dopaminergic neuron loss or notable changes in striatal dopamine levels at 12 months of age (18). Interestingly, recent findings suggest that tau missorting in these mice may contribute to the development of dementia in familial A53T α-synuclein-linked PD cases (76). This observation underscores the potential interplay between α-synucleinopathy and tauopathy, providing valuable insights into the broader neurodegenerative mechanisms underlying PD and related disorders.

2.3.2.2 Parkin-Based Genetic Model

Mutations in the Parkin gene have been strongly associated with autosomal recessive juvenile Parkinsonism, a form of early-onset PD. The Parkin protein functions as an E3 ubiquitin ligase and plays a pivotal role in maintaining neuronal health by targeting damaged mitochondria for clearance via mitophagy, a specialized form of autophagy (79).

Loss-of-function mutations in the Parkin gene result in a dysfunctional Parkin protein, which compromises mitochondrial quality control. This dysfunction leads to the accumulation of damaged mitochondria, increased oxidative stress, and a heightened vulnerability of nigral dopaminergic neurons to degeneration (80–82). Conversely, several studies have demonstrated that Parkin overexpression can exert neuroprotective effects. In rodent models of PD, overexpression of the *parkin* gene conferred resistance to 6-OHDA-induced lesions and α-synuclein overexpression, effectively preserving dopaminergic neurons in the substantia nigra (83–85). In Parkin-deficient mice, motor impairments have been observed through behavioral assays such as beam walking and adhesive removal tests, along with reduced synaptic excitability in the striatum, an area critical for motor function. However, these models do not typically display overt dopaminergic neuronal loss in the substantia nigra, suggesting that behavioral and synaptic alterations may precede or occur independently of neurodegeneration in this model (86). Interestingly, Parkin knockout rats have shown limited phenotypic changes compared to their mouse counterparts. These rats did not exhibit significant behavioral, neurochemical, or pathological differences from wild-type controls (87), and notably, Lewy body-like pathology or other α-synuclein-immunoreactive aggregates were absent (88). Despite the lack of classic PD pathology, Parkin knockout rats remain a valuable model for proteomic and mechanistic studies. They facilitate the identification of Parkin substrates and help uncover the broader implications of Parkin dysfunction in PD pathogenesis (89,90).

2.3.2.3 DJ-1-Based Genetic Model

The DJ-1 protein, also known as PD protein 7 and encoded by the PARK7 gene, plays a crucial role in protecting neurons from oxidative stress and maintaining mitochondrial homeostasis. Mutations or deletion of this gene are linked to autosomal recessive early-onset PD. To study its function and pathological relevance, researchers have developed both DJ-1 knockout and DJ-1 mutant models, which provide valuable insights into the disease mechanisms and therapeutic strategies.

2.3.2.3.1 DJ-1 Knockout Models

DJ-1 knockout mice are generated using homologous recombination in embryonic stem cells, resulting in the complete absence of DJ-1 protein expression. These models are instrumental in elucidating the physiological role of DJ-1 in neuronal survival and cellular defense mechanisms. DJ-1 Knockout mice commonly exhibit reduced spontaneous locomotion, motor coordination impairments, and heightened susceptibility to neurotoxic agents. Behavioral assays, including the rotarod test and open field test, demonstrate motor deficits that resemble Parkinsonian features. These models consistently show increased oxidative stress, impaired mitochondrial dynamics, and disrupted energy metabolism. Biochemical evaluations reveal deficits in ATP production, elevated levels of ROS, and abnormal dopamine metabolism, reflecting early pathological events implicated in neurodegeneration.

2.3.2.3.2 DJ-1 Mutant Models

In addition to knockout models, point mutation models of DJ-1 have been developed to study familial forms of PD. These models carry pathogenic mutations such as M146L and L166P, which are associated with functional loss or destabilization of the DJ-1 protein. Transgenic mice expressing mutant DJ-1 proteins exhibit dopaminergic neuron loss in the substantia nigra, along with behavioral impairments reflective of progressive PD. These alterations are accompanied by neurochemical imbalances in the nigrostriatal dopamine system.

Behavioral and cognitive assessments, combined with neurochemical profiling, have identified both subtle and marked deficits in motor and non-motor domains. These findings mirror key clinical features of PD and provide a robust platform for mechanistic exploration.

Collectively, DJ-1 knockout and mutant models have significantly contributed to our understanding of the molecular underpinnings of PD, especially those related to oxidative stress, mitochondrial dysfunction, and dopaminergic vulnerability. These models continue to serve as vital tools for both pathophysiological research and preclinical drug development (91,92).

2.3.2.4 PINK1-Based Genetic Model

The PINK1 gene (PTEN-induced putative kinase 1) plays a critical role in mitochondrial quality control by promoting the removal of dysfunctional mitochondria via mitophagy. Genetic models targeting PINK1 have been instrumental in elucidating the pathophysiological mechanisms underlying early-onset PD.

2.3.2.4.1 PINK1 Knockout Models

PINK1 Knockout models are typically generated using either CRISPR/Cas9 gene-editing technology or homologous recombination techniques. The absence of PINK1 disrupts the mitophagy pathway, leading to the accumulation of damaged mitochondria and increased susceptibility to neuronal death. PINK1 Knockout mice display a spectrum of PD-like phenotypes, including motor coordination deficits, reduced locomotor activity, altered gait patterns, and increased sensitivity to neurotoxins such as MPTP. Histological analysis often reveals neurodegeneration in the substantia nigra. These models also demonstrate elevated levels of ROS and widespread oxidative stress. Additionally, abnormal mitochondrial morphology, including swelling and fragmentation, is frequently observed through electron microscopy, reflecting impaired mitochondrial dynamics and homeostasis.

2.3.2.4.2 Mutant PINK1 Models

In addition to knockout models, transgenic mice expressing disease-associated PINK1 mutations, such as G309D, offer valuable insights into the mechanisms of mitochondrial dysfunction in familial PD. Mice carrying the G309D missense mutation exhibit progressive neurodegeneration, impaired motor behavior, and increased oxidative stress. These features closely mimic the clinical and pathological progression seen in early-onset PD linked to PINK1 mutations. Mutant PINK1 models are widely used in preclinical drug screening, particularly for compounds aimed at enhancing mitochondrial function, promoting mitophagy, or activating compensatory pathways that protect against neuronal death. These models serve as essential tools for developing and evaluating potential disease-modifying therapies (93–95).

2.3.2.5 LRRK2-Based Genetic Models

Mutations in the leucine-rich repeat kinase 2 (LRRK2) gene are among the most significant genetic contributors to both familial and sporadic forms of PD. The most common pathogenic variant, G2019S, is associated with autosomal dominant PD and is frequently used in experimental models to study disease mechanisms. Several strategies have been employed to model LRRK2-associated PD, including knock-in, knock-out, and viral vector-mediated approaches.

2.3.2.5.1 LRRK2 Knock-in Models

Knock-in models are developed by introducing specific human LRRK2 mutations, particularly G2019S, into the endogenous mouse genome using CRISPR/Cas9 genome editing or viral vector-mediated gene transfer. These models are instrumental in studying the pathogenic role of LRRK2 mutations under physiological expression levels. Human LRRK2 variants are inserted into the mouse genome to mimic the genetic context of PD patients. These models maintain endogenous regulatory elements, providing a more accurate reflection of disease progression. LRRK2 knock-in mice typically develop age-dependent motor deficits, dopaminergic neuron loss, and increased α-synuclein aggregation, particularly in the substantia nigra. Furthermore, these models have provided key insights into autophagy dysregulation, synaptic dysfunction, and increased kinase activity, all of which are central to LRRK2-mediated neurotoxicity. Importantly, LRRK2 kinase inhibitors have been shown to attenuate neuronal damage, underscoring the therapeutic relevance of this pathway.

2.3.2.5.2 LRRK2 Knockout Models

Knock-out models lacking functional LRRK2 help to elucidate the consequences of loss-of-function mutations and the gene's role in neuronal maintenance and immune regulation. While some LRRK2 knockout mice display mild or no motor deficits, others show subtle impairments in coordination and dopamine metabolism, suggesting strain-dependent effects. These models have also revealed LRRK2's involvement in microglial activation, neuroinflammation, and lysosomal function.

2.3.2.5.3 Viral Vector Models of LRRK2 Overexpression

Adeno-associated viruses (AAVs) and other viral delivery systems are employed to express LRRK2 mutations, such as G2019S, directly in specific brain regions. Stereotactic injection of AAV vectors into the substantia nigra enables localized overexpression of mutant LRRK2, leading to dopaminergic neuronal loss, α-synuclein pathology, and motor impairments, closely mimicking key features of human PD (96–98).

REFERENCES

1. Chaudhuri KR, Healy DG, Schapira AHV. Non-motor symptoms of Parkinson's disease: diagnosis and management. Lancet Neurol [Internet]. 2006 Mar 1 [cited 2024 Jul 9];5(3):235–45. Available from: http://www.thelancet.com/article/S1474442206703738/fulltext
2. Lohr KM, Bernstein AI, Stout KA, Dunn AR, Lazo CR, Alter SP, et al. Increased vesicular monoamine transporter enhances dopamine release and opposes Parkinson disease-related neurodegeneration in vivo. Proc Natl Acad Sci USA. 2014 Jul;111(27):9977–82.
3. Duty S, Jenner P. Animal models of Parkinson's disease: a source of novel treatments and clues to the cause of the disease. Br J Pharmacol. 2011 Oct;164(4):1357–91.
4. Dauer W, Przedborski S. Parkinson's disease: mechanisms and models. Neuron. 2003 Sep 11; 39(6):889–909.
5. Schapira AH V, Chaudhuri KR, Jenner P. Non-motor features of Parkinson disease. Nat Rev Neurosci. 2017 Jul;18(7):435–50.
6. Pingale T, Gupta GL. Classic and evolving animal models in Parkinson's disease. Pharmacol Biochem Behav. 2020 Dec 1;199.
7. de Freitas CM, Busanello A, Schaffer LF, Peroza LR, Krum BN, Leal CQ, et al. Behavioral and neurochemical effects induced by reserpine in mice. Psychopharmacology (Berl). 2016 Feb;233(3):457–67.
8. Atack JR, Shook BC, Rassnick S, Jackson PF, Rhodes K, Drinkenburg WH, et al. JNJ-40255293, a novel adenosine A2A/A1 antagonist with efficacy in preclinical models of Parkinson's disease. ACS Chem Neurosci. 2014 Oct;5(10):1005–19.
9. Engers DW, Blobaum AL, Gogliotti RD, Cheung YY, Salovich JM, Garcia-Barrantes PM, et al. Discovery, synthesis, and preclinical characterization of N-(3-chloro-4-fluorophenyl)-1H-pyrazolo[4,3-b]pyridin-3-amine (VU0418506), a novel positive allosteric modulator of the metabotropic glutamate receptor 4 (mGlu4). ACS Chem Neurosci. 2016 Sep;7(9):1192–200.
10. Waku I, Magalhães MS, Alves CO, de Oliveira AR. Haloperidol-induced catalepsy as an animal model for parkinsonism: A systematic review of experimental studies. Eur J Neurosci. 2021 Jun;53(11):3743–67.
11. Yael D, Zeef DH, Sand D, Moran A, Katz DB, Cohen D, et al. Haloperidol-induced changes in neuronal activity in the striatum of the freely moving rat. Front Syst Neurosci. 2013;7:110.
12. Kumari N, Agrawal S, Luthra PM. Pharmacological assessments of potent A(2A) receptor antagonist IDPU (1-(7-imino-3-propyl-2,3-dihydrothiazolo[4,5-d]pyrimidin-6(7H)-yl)urea) in rodent model of haloperidol induced Parkinson like symptoms. Neurosci Lett. 2017 Apr;647:53–60.
13. Arora D, Mudgal J, Nampoothiri M, Mallik SB, Kinra M, Hall S, et al. Interplay between adenosine receptor antagonist and cyclooxygenase inhibitor in haloperidol-induced extrapyramidal effects in mice. Metab Brain Dis. 2018 Aug;33(4):1045–51.
14. Ungerstedt U, Arbuthnott GW. Quantitative recording of rotational behavior in rats after 6-hydroxydopamine lesions of the nigrostriatal dopamine system. Brain Res [Internet]. 1970 Dec 18 [cited 2024 Jan 15];24(3):485–93. Available from: https://pubmed.ncbi.nlm.nih.gov/5494536/
15. Chotibut T, Apple DM, Jefferis R, Salvatore MF. Dopamine transporter loss in 6-OHDA Parkinson's model is unmet by parallel reduction in dopamine uptake. PLoS One [Internet]. 2012 Dec 29 [cited 2024 Jan 20];7(12):e52322. Available from: https://journals.plos.org/plosone/article?id=10.1371/journal.pone.0052322
16. Konnova EA, Swanberg M. Animal models of Parkinson's disease. Exon Publications [Internet]. 2018 Dec 21 [cited 2024 Jan 20];83–106. Available from: https://exonpublications.com/index.php/exon/article/view/190/346

17. Bové J, Perier C. Neurotoxin-based models of Parkinson's disease. Neuroscience [Internet]. 2012 Jun 1 [cited 2024 Jan 20];211:51–76. Available from: http://www.ibroneuroscience.org/article/S0306452211012668/fulltext
18. Dovonou A, Bolduc C, Soto Linan V, Gora C, Peralta MR, Lévesque M. Animal models of Parkinson's disease: bridging the gap between disease hallmarks and research questions. Transl Neurodegener [Internet]. 2023 Jul 19 [cited 2024 Jan 20];12(1):36–36. Available from: https://europepmc.org/articles/PMC10354932
19. Hadipour E, Fereidoni M, Tayarani-Najaran Z. Betanin attenuates oxidative stress induced by 6-OHDA in PC12 cells via SAPK/JNK and PI3 K pathways. Neurochem Res [Internet]. 2020 [cited 2024 Jan 27]; 45:395–403. Available from: https://doi.org/10.1007/s11064-019-02927-w
20. Wang SF, Liu LF, Wu MY, Cai CZ, Su H, Tan J, et al. Baicalein prevents 6-OHDA/ascorbic acid-induced calcium-dependent dopaminergic neuronal cell death. [cited 2024 Jan 27]; Available from: www.nature.com/scientificreports/
21. Bernstein AI, Garrison SP, Zambetti GP, O'malley KL. 6-OHDA generated ROS induces DNA damage and p53- and PUMA-dependent cell death. Mol Neurodegener [Internet]. 2011 [cited 2024 Jan 27];6:2. Available from: http://www.molecularneurodegeneration.com/content/6/1/2
22. Wang L, Xu S, Xu X, Chan P. (–)-Epigallocatechin-3-gallate protects SH-SY5Y cells against 6-OHDA-induced cell death through STAT3 activation. J Alzheimer's Dis. 2009 Jan 1;17(2):295–304.
23. Tian Y, Lu J, Hao X, Li H, Zhang G, Liu X, et al. FTH1 inhibits ferroptosis through ferritinophagy in the 6-OHDA model of Parkinson's disease. [cited 2024 Jan 27]; Available from: https://doi.org/10.1007/s13311-020-00929-z
24. Lu X, Kim-Han JS, Harmon S, Sakiyama-Elbert SE, O'Malley KL. The Parkinsonian mimetic, 6-OHDA, impairs axonal transport in dopaminergic axons. Mol Neurodegener [Internet]. 2014 May 3 [cited 2024 Jan 28];9(1):1–11. Available from: https://molecularneurodegeneration.biomedcentral.com/articles/10.1186/1750-1326-9-17
25. Curtius HC, Wolfensberger M, Steinmann B, Redweik U, Siegfried J. Mass fragmentography of dopamine and 6-hydroxydopamine. Application to the determination of dopamine in human brain biopsies from the caudate nucleus. J Chromatogr A. 1974 Nov 6;99(C):529–40.
26. Andrew R, Watson DG, Best SA, Midgley JM, Wenlong H, Petty RKH. The determination of hydroxydopamines and other trace amines in the urine of Parkinsonian patients and normal controls. Neurochem Res [Internet]. 1993 Nov [cited 2024 Jan 28];18(11):1175–7. Available from: https://link.springer.com/article/10.1007/BF00978370
27. Van Den Berge N, Ulusoy A. Animal models of brain-first and body-first Parkinson's disease. 2021 [cited 2024 Jan 28]; Available from: https://doi.org/10.1016/j.nbd.2021.105599
28. Davis GC, Williams AC, Markey SP, Ebert MH, Caine ED, Reichert CM, et al. Chronic Parkinsonism secondary to intravenous injection of meperidine analogues. Psychiatry Res. 1979 Dec;1(3):249–54.
29. Langston JW, Ballard P, Tetrud JW, Irwin I. Chronic Parkinsonism in humans due to a product of meperidine-analog synthesis. Science. 1983 Feb;219(4587):979–80.
30. Langston JW, Ballard PAJ. Parkinson's disease in a chemist working with 1-methyl-4-phenyl-1,2,5,6-tetrahydropyridine. New Engl J Med. United States; 1983;309:310.
31. Blesa J, Przedborski S. Parkinson's disease: animal models and dopaminergic cell vulnerability. Front Neuroanat. 2014;8:155.
32. Del Zompo M, Piccardi MP, Ruiu S, Quartu M, Gessa GL, Vaccari A. Selective MPP+ uptake into synaptic dopamine vesicles: possible involvement in MPTP neurotoxicity. Br J Pharmacol. 1993 Jun;109(2): 411–4.
33. Bezard E, Gross CE, Fournier MC, Dovero S, Bloch B, Jaber M. Absence of MPTP-induced neuronal death in mice lacking the dopamine transporter. Exp Neurol. 1999 Feb;155(2):268–73.
34. Fornai F, Schlüter OM, Lenzi P, Gesi M, Ruffoli R, Ferrucci M, et al. Parkinson-like syndrome induced by continuous MPTP infusion: convergent roles of the ubiquitin-proteasome system and alpha-synuclein. Proc Natl Acad Sci USA. 2005 Mar;102(9):3413–8.
35. Tillerson JL, Caudle WM, Reverón ME, Miller GW. Detection of behavioral impairments correlated to neurochemical deficits in mice treated with moderate doses of 1-methyl-4-phenyl-1,2,3,6-tetrahydropyridine. Exp Neurol. 2002 Nov;178(1):80–90.
36. Schintu N, Frau L, Ibba M, Caboni P, Garau A, Carboni E, et al. PPAR-gamma-mediated neuroprotection in a chronic mouse model of Parkinson's disease. Eur J Neurosci. 2009 Mar;29(5):954–63.
37. Laloux C, Derambure P, Kreisler A, Houdayer E, Bruezière S, Bordet R, et al. MPTP-treated mice: long-lasting loss of nigral TH-ir neurons but not paradoxical sleep alterations. Exp Brain Res. 2008 Apr;186(4):635–42.

38. Laloux C, Derambure P, Houdayer E, Jacquesson JM, Bordet R, Destée A, et al. Effect of dopaminergic substances on sleep/wakefulness in saline- and MPTP-treated mice. J Sleep Res. 2008 Mar;17(1):101–10.
39. Anderson G, Noorian AR, Taylor G, Anitha M, Bernhard D, Srinivasan S, et al. Loss of enteric dopaminergic neurons and associated changes in colon motility in an MPTP mouse model of Parkinson's disease. Exp Neurol. 2007 Sep;207(1):4–12.
40. Vucković MG, Wood RI, Holschneider DP, Abernathy A, Togasaki DM, Smith A, et al. Memory, mood, dopamine, and serotonin in the 1-methyl-4-phenyl-1,2,3,6-tetrahydropyridine-lesioned mouse model of basal ganglia injury. Neurobiol Dis. 2008 Nov;32(2):319–27.
41. Schmidt N, Ferger B. Neurochemical findings in the MPTP model of Parkinson's disease. J Neural Transm (Vienna). 2001;108(11):1263–82.
42. Schober A. Classic toxin-induced animal models of Parkinson's disease: 6-OHDA and MPTP. Cell Tissue Res. 2004 Oct;318(1):215–24.
43. Riachi NJ, LaManna JC, Harik SI. Entry of 1-methyl-4-phenyl-1,2,3,6-tetrahydropyridine into the rat brain. J Pharmacol Exp Ther. 1989 Jun;249(3):744–8.
44. Riachi NJ, Harik SI, Kalaria RN, Sayre LM. On the mechanisms underlying 1-methyl-4-phenyl-1,2,3,6-tetrahydropyridine neurotoxicity. II. Susceptibility among mammalian species correlates with the toxin's metabolic patterns in brain microvessels and liver. J Pharmacol Exp Ther. 1988 Feb;244(2):443–8.
45. Cenci MA, Whishaw IQ, Schallert T. Animal models of neurological deficits: how relevant is the rat? Nat Rev Neurosci. 2002 Jul;3(7):574–9.
46. Przedborski S, Jackson-Lewis V, Naini AB, Jakowec M, Petzinger G, Miller R, et al. The parkinsonian toxin 1-methyl-4-phenyl-1,2,3,6-tetrahydropyridine (MPTP): a technical review of its utility and safety. J Neurochem. 2001 Mar;76(5):1265–74.
47. Sedelis M, Schwarting RK, Huston JP. Behavioral phenotyping of the MPTP mouse model of Parkinson's disease. Behav Brain Res. 2001 Nov;125(1–2):109–25.
48. Jackson-Lewis V, Przedborski S. Protocol for the MPTP mouse model of Parkinson's disease. Nat Protoc. 2007;2(1):141–51.
49. Perier C, Bové J, Wu DC, Dehay B, Choi DK, Jackson-Lewis V, et al. Two molecular pathways initiate mitochondria-dependent dopaminergic neurodegeneration in experimental Parkinson's disease. Proc Natl Acad Sci USA. 2007 May;104(19):8161–6.
50. Prediger RDS, Aguiar ASJ, Rojas-Mayorquin AE, Figueiredo CP, Matheus FC, Ginestet L, et al. Single intranasal administration of 1-methyl-4-phenyl-1,2,3,6-tetrahydropyridine in C57BL/6 mice models early preclinical phase of Parkinson's disease. Neurotox Res. 2010 Feb;17(2):114–29.
51. Prediger RDS, Aguiar ASJ, Moreira ELG, Matheus FC, Castro AA, Walz R, et al. The intranasal administration of 1-methyl-4-phenyl-1,2,3,6-tetrahydropyridine (MPTP): a new rodent model to test palliative and neuroprotective agents for Parkinson's disease. Curr Pharm Des. 2011;17(5):489–507.
52. Lee E, Hwang I, Park S, Hong S, Hwang B, Cho Y, et al. MPTP-driven NLRP3 inflammasome activation in microglia plays a central role in dopaminergic neurodegeneration. Cell Death Differ. 2019 Jan;26(2):213–28.
53. Smeyne M, Jiao Y, Shepherd KR, Smeyne RJ. Glia cell number modulates sensitivity to MPTP in mice. Glia. 2005 Nov;52(2):144–52.
54. Muthane U, Ramsay KA, Jiang H, Jackson-Lewis V, Donaldson D, Fernando S, et al. Differences in nigral neuron number and sensitivity to 1-methyl-4-phenyl-1,2,3,6-tetrahydropyridine in C57/bl and CD-1 mice. Exp Neurol. 1994 Apr;126(2):195–204.
55. Yin DX, Toyoda H, Nozaki K, Satoh K, Katagiri A, Adachi K, et al. Taste impairments in a Parkinson's disease model featuring intranasal rotenone administration in mice. J Parkinsons Dis. 2022;12(6):1863–80.
56. Sato H, Satoh K, Nozaki K, Yugawa M, Kato T, Toyoda H, et al. Reduced menthol sensitivity in a prodromal Parkinson's disease model induced by intranasal rotenone treatment. Front Cell Neurosci. 2024;18:1345651.
57. de Souza Nascimento T, Oliveira AV, de Oliveira Bélem M, Bezerra JR, do Carmo MRS, da Silva ME, et al. The rotenone-induced sporadic Parkinsonism model: timeline of motor and non-motor features. Eur J Neurosci. 2025 Feb;61(3):e16669.
58. Troshev D, Voronkov D, Pavlova A, Abaimov D, Latanov A, Fedorova T, et al. Time course of neurobehavioral disruptions and regional brain metabolism changes in the rotenone mice model of Parkinson's disease. Biomedicines. 2022 Feb;10(2).
59. Tian L, Tang P, Liu J, Liu Y, Hou L, Zhao J, et al. Microglial gp91phox-mediated neuroinflammation and ferroptosis contributes to learning and memory deficits in rotenone-treated mice. Free Radic Biol Med. 2024 Aug;220:56–66.

60. Gupta V, Prasad S. Differential alterations in the expression of AMPA receptor and its trafficking proteins in the hippocampus are associated with recognition memory impairment in the rotenone-Parkinson's disease mouse model: neuroprotective role of *Bacopa monnieri* extract. Mol Neurobiol. 2025 Feb;62(2):2086–104.
61. Landau R, Halperin R, Sullivan P, Zibly Z, Leibowitz A, Goldstein DS, et al. The rat rotenone model reproduces the abnormal pattern of central catecholamine metabolism found in Parkinson's disease. Dis Model Mech. 2022 Jan;15(1).
62. Guo Z, Ruan Z, Zhang D, Liu X, Hou L, Wang Q. Rotenone impairs learning and memory in mice through microglia-mediated blood brain barrier disruption and neuronal apoptosis. Chemosphere. 2022 Mar;291(Pt 2):132982.
63. Zaman V, Matzelle D, Banik NL, Haque A. Dysregulation of metabolic peptides precedes hyperinsulinemia and inflammation following exposure to rotenone in rats. Cells. 2025 Jan;14(2).
64. Fang X, Liu S, Muhammad B, Zheng M, Ge X, Xu Y, et al. Gut microbiota dysbiosis contributes to α-synuclein-related pathology associated with C/EBPβ/AEP signaling activation in a mouse model of Parkinson's disease. Neural Regen Res. 2024 Sep;19(9):2081–8.
65. Sharma M, Sharma N, Khairnar A. Intranasal rotenone induces alpha-synuclein accumulation, neuroinflammation and dopaminergic neurodegeneration in middle-aged mice. Neurochem Res. 2023 May;48(5):1543–60.
66. Alam M, Mayerhofer A, Schmidt WJ. The neurobehavioral changes induced by bilateral rotenone lesion in medial forebrain bundle of rats are reversed by L-DOPA. Behav Brain Res [Internet]. 2004;151(1):117–24. Available from: https://www.sciencedirect.com/science/article/pii/S0166432803003000
67. Xiong N, Huang J, Zhang Z, Zhang Z, Xiong J, Liu X, et al. Stereotaxical infusion of rotenone: a reliable rodent model for Parkinson's disease. PLoS One. 2009 Nov;4(11):e7878.
68. Talpade DJ, Greene JG, Higgins DSJ, Greenamyre JT. In vivo labeling of mitochondrial complex I (NADH:ubiquinone oxidoreductase) in rat brain using [(3)H]dihydrorotenone. J Neurochem. 2000 Dec;75(6):2611–21.
69. Huang M, Wang B, Li X, Fu C, Wang C, Kang X. α-Synuclein: a multifunctional player in exocytosis, endocytosis, and vesicle recycling. Front Neurosci. 2019;13:28.
70. Xu L, Pu J. Alpha-synuclein in Parkinson's disease: from pathogenetic dysfunction to potential clinical application. Parkinsons Dis. 2016;2016:1720621.
71. Liu H, Koros C, Strohäker T, Schulte C, Bozi M, Varvaresos S, et al. A novel SNCA A30G mutation causes familial Parkinson's disease. Mov Disord. 2021 Jul;36(7):1624–33.
72. Emmer KL, Waxman EA, Covy JP, Giasson BI. E46K human alpha-synuclein transgenic mice develop Lewy-like and tau pathology associated with age-dependent, detrimental motor impairment. J Biol Chem. 2011 Oct;286(40):35104–18.
73. Cannon JR, Geghman KD, Tapias V, Sew T, Dail MK, Li C, et al. Expression of human E46K-mutated α-synuclein in BAC-transgenic rats replicates early-stage Parkinson's disease features and enhances vulnerability to mitochondrial impairment. Exp Neurol. 2013 Feb;240:44–56.
74. Piltonen M, Savolainen M, Patrikainen S, Baekelandt V, Myöhänen TT, Männistö PT. Comparison of motor performance, brain biochemistry and histology of two A30P α-synuclein transgenic mouse strains. Neuroscience. 2013 Feb;231:157–68.
75. Ekmark-Lewén S, Lindström V, Gumucio A, Ihse E, Behere A, Kahle PJ, et al. Early fine motor impairment and behavioral dysfunction in (Thy-1)-h[A30P] alpha-synuclein mice. Brain Behav. 2018 Mar;8(3):e00915.
76. Teravskis PJ, Covelo A, Miller EC, Singh B, Martell-Martínez HA, Benneyworth MA, et al. A53T mutant alpha-synuclein induces tau-dependent postsynaptic impairment independently of neurodegenerative changes. J Neurosci. 2018 Nov;38(45):9754–67.
77. Mougenot ALJ, Bencsik A, Nicot S, Vulin J, Morignat E, Verchère J, et al. Transmission of prion strains in a transgenic mouse model overexpressing human A53T mutated α-synuclein. J Neuropathol Exp Neurol. 2011 May;70(5):377–85.
78. Giasson BI, Duda JE, Quinn SM, Zhang B, Trojanowski JQ, Lee VMY. Neuronal alpha-synucleinopathy with severe movement disorder in mice expressing A53T human alpha-synuclein. Neuron. 2002 May;34(4):521–33.
79. Zhang Y, Gao J, Chung KK, Huang H, Dawson VL, Dawson TM. Parkin functions as an E2-dependent ubiquitin- protein ligase and promotes the degradation of the synaptic vesicle-associated protein, CDCrel-1. Proc Natl Acad Sci USA. 2000 Nov;97(24):13354–9.
80. Cookson MR. DJ-1, PINK1, and their effects on mitochondrial pathways. Mov Disord. 2010;25 Suppl 1(Suppl 1):S44–8.

81. Trempe JF, Fon EA. Structure and function of Parkin, PINK1, and DJ-1, the three musketeers of neuroprotection. Front Neurol. 2013;4:38.
82. Kalinderi K, Bostantjopoulou S, Fidani L. The genetic background of Parkinson's disease: current progress and future prospects. Acta Neurol Scand. 2016 Nov;134(5):314–26.
83. Lo Bianco C, Schneider BL, Bauer M, Sajadi A, Brice A, Iwatsubo T, et al. Lentiviral vector delivery of parkin prevents dopaminergic degeneration in an alpha-synuclein rat model of Parkinson's disease. Proc Natl Acad Sci U S A. 2004 Dec;101(50):17510–5.
84. Yamada M, Mizuno Y, Mochizuki H. Parkin gene therapy for alpha-synucleinopathy: a rat model of Parkinson's disease. Hum Gene Ther. 2005 Feb;16(2):262–70.
85. Vercammen L, Van der Perren A, Vaudano E, Gijsbers R, Debyser Z, Van den Haute C, et al. Parkin protects against neurotoxicity in the 6-hydroxydopamine rat model for Parkinson's disease. Mol Ther. 2006 Nov;14(5):716–23.
86. Goldberg MS, Fleming SM, Palacino JJ, Cepeda C, Lam HA, Bhatnagar A, et al. Parkin-deficient mice exhibit nigrostriatal deficits but not loss of dopaminergic neurons. J Biol Chem. 2003 Oct;278(44):43628–35.
87. Dave KD, De Silva S, Sheth NP, Ramboz S, Beck MJ, Quang C, et al. Phenotypic characterization of recessive gene knockout rat models of Parkinson's disease. Neurobiol Dis. 2014 Oct;70:190–203.
88. Doherty KM, Hardy J. Parkin disease and the Lewy body conundrum. Mov Disord. 2013 Jun; 28(6):702–4.
89. Stauch KL, Villeneuve LM, Purnell PR, Pandey S, Guda C, Fox HS. SWATH-MS proteome profiling data comparison of DJ-1, Parkin, and PINK1 knockout rat striatal mitochondria. Data Brief. 2016 Dec;9:589–93.
90. Kurup PK, Xu J, Videira RA, Ononenyi C, Baltazar G, Lombroso PJ, et al. STEP61 is a substrate of the E3 ligase parkin and is upregulated in Parkinson's disease. Proc Natl Acad Sci USA. 2015 Jan;112(4):1202–7.
91. Chen P, DeWitt MR, Bornhorst J, Soares FA, Mukhopadhyay S, Bowman AB, et al. Age- And manganese-dependent modulation of dopaminergic phenotypes in a *C. elegans* DJ-1 genetic model of Parkinson's disease. Metallomics [Internet]. 2015 Mar 1 [cited 2025 May 4];7(2):289–98. Available from: https://pubmed.ncbi.nlm.nih.gov/25531510/
92. Arboleya S, Watkins C, Stanton C, Ross RP. Gut bifidobacteria populations in human health and aging. Front Microbiol [Internet]. 2016 Aug 19 [cited 2025 May 4];7(AUG). Available from: https://pubmed.ncbi.nlm.nih.gov/27594848/
93. Han R, Liu Y, Li S, Li XJ, Yang W. PINK1-PRKN mediated mitophagy: differences between in vitro and in vivo models. Autophagy [Internet]. 2023 [cited 2025 May 4];19(5):1396–405. Available from: https://pubmed.ncbi.nlm.nih.gov/36282767/
94. QuinnPMJ, Moreira PI, Ambrósio AF, Alves CH. PINK1/PARKIN signalling in neurodegeneration and neuroinflammation. Acta Neuropathol Commun [Internet]. 2020 Nov 9 [cited 2025 May 4];8(1):189. Available from: https://pubmed.ncbi.nlm.nih.gov/33168089/
95. VizzielloM, Borellini L, Franco G, Ardolino G. Disruption of mitochondrial homeostasis: the role of pink1 in Parkinson's disease. Cells [Internet]. 2021 Nov 1 [cited 2025 May 4];10(11). Available from: https://pubmed.ncbi.nlm.nih.gov/34831247/
96. Usmani A, Shavarebi F, Hiniker A. The cell biology of LRRK2 in Parkinson's disease. Mol Cell Biol [Internet]. 2021 May 1 [cited 2025 May 4];41(5). Available from: https://pubmed.ncbi.nlm.nih.gov/33526455/
97. Deniston CK, Salogiannis J, Mathea S, Snead DM, Lahiri I, Matyszewski M, et al. Structure of LRRK2 in Parkinson's disease and model for microtubule interaction. Nature [Internet]. 2020 Dec 10 [cited 2025 May 4];588(7837):344–9. Available from: https://pubmed.ncbi.nlm.nih.gov/32814344/
98. Liang D, Liu H, Jin R, Feng R, Wang J, Qin C, et al. Escherichia coli triggers α-synuclein pathology in the LRRK2 transgenic mouse model of PD. Gut Microbes [Internet]. 2023 [cited 2025 May 4];15(2). Available from: https://www.tandfonline.com/doi/pdf/10.1080/19490976.2023.2276296?needAccess=true

3 Rodent Models of Amyotrophic Lateral Sclerosis

Fatemeh Mokhtari Naseri

3.1 INTRODUCTION

Amyotrophic lateral sclerosis (ALS) is a complex and fatal neurodegenerative disease characterized by the progressive degeneration of both upper motor neurons (UMNs) and lower motor neurons (LMNs). This results in severe motor dysfunction, including muscle weakness, spasticity, dysphagia, and eventual paralysis. ALS manifests heterogeneously, with some patients experiencing spinal-onset ALS, marked by limb weakness, while others present with bulbar-onset ALS, characterized by difficulties in speech and swallowing.

Although the exact cause of ALS remains unknown in most cases, about 10% of cases are familial (fALS), linked to mutations in genes with diverse cellular functions, including those in non-motor cells. Notably, the chromosome 9 open reading frame 72 (C9orf72) hexanucleotide repeat expansion is the most common genetic cause of both ALS and frontotemporal dementia (FTD), highlighting the overlap between motor and cognitive pathology. In fact, up to 50% of ALS patients develop cognitive and/or behavioral impairments, and approximately 13% present with behavioral variant FTD (bvFTD). This has led to a paradigm shift in ALS research, redefining it as a multisystem neurodegenerative disorder rather than a purely neuromuscular disease. Rodent models have played a pivotal role in unraveling the pathophysiology of ALS and testing potential therapeutic strategies. Over the years, transgenic, toxin-induced, and viral vector-based models have been developed to recapitulate key aspects of ALS pathology.

This chapter begins with an overview of ALS pathology, highlighting the key features of the disease, including motor neuron degeneration, protein aggregation, and the involvement of glial cells. Following this, a detailed discussion of rodent models of ALS will be provided, exploring the various models used to replicate different aspects of ALS pathology, their strengths and limitations, and their contributions to advancing our understanding of the disease and testing potential therapies.

3.2 ALS NEUROPATHOLOGY

ALS is a multifaceted neurodegenerative disorder marked by the accumulation of ubiquitinated proteinaceous inclusions within motor neurons. This neuropathological hallmark is shared with other neurodegenerative diseases, including Alzheimer's disease and Parkinson's disease. Despite extensive research, the precise biological mechanisms underlying the formation of these inclusions remain incompletely understood. In most ALS cases, transactive response DNA-binding protein 43 (TDP-43) is the main component of cytoplasmic inclusions, although mutations in its encoding gene, TARDBP, account for only a small fraction of familial cases. Notably, approximately 97% of ALS patients exhibit TDP-43 proteinopathy, characterized by nuclear depletion and cytoplasmic aggregation of TDP-43, often presenting as skein-like or compact inclusions (1).

Distinct subtypes of ALS may feature additional protein aggregates, such as neurofilamentous hyaline conglomerates or misfolded superoxide dismutase 1 (SOD1) deposits, the latter being characteristic of cases with *SOD1* mutations. In ALS associated with C9orf72 mutations, inclusions may extend beyond the motor system, further complicating the disease pathology. While protein inclusions are pathological hallmarks of ALS, it is increasingly hypothesized that the toxic species

DOI: 10.1201/9781032676524-4

driving disease progression are the high-molecular-weight protein complexes that precede inclusion formation. These complexes may contribute to cell-to-cell propagation of pathology via prion-like mechanisms, similar to those observed in tauopathies and synucleinopathies (2).

Gross and microscopic pathological features of ALS include skeletal muscle atrophy, pallor and sclerosis of the pyramidal tracts, motor cortex atrophy, and thinning of the hypoglossal nerves and ventral spinal roots. On microscopic examination, ALS is characterized by the loss of more than 50% of spinal motor neurons, accompanied by diffuse astrocytic gliosis and microglial infiltration within both the gray and white matter. The motor cortex typically shows astrogliosis with variable depletion of UMNs. Affected skeletal muscle exhibits classic features of denervation and reinnervation, such as fiber-type grouping and angular atrophic fibers. The classic ALS phenotype presents with a combination of muscle atrophy, fasciculations, hyporeflexia, hypotonia, stiffness, clumsiness, and painful flexor spasms (3,4). Other notable variants include the following:

- *Progressive Muscular Atrophy (PMA)*: A LMN-dominant form.
- *UMN-Dominant ALS*: Typically shows slower progression than the classic phenotype.
- *Respiratory Dysfunction*: Characterized by ventilatory impairment and symptoms such as orthopnea, morning headaches, and exertional dyspnea. As weakness progresses, tachypnea, accessory muscle use, and paradoxical breathing may occur.
- *Neurobehavioral Dysfunction*: FTD co-occurs in about 15% of ALS patients, while milder behavioral and cognitive impairments affect up to 75%, correlating with reduced survival.
- *Pain*: Reported in approximately half of ALS patients, pain is multifactorial and often underrecognized and undertreated.

3.3 RODENT MODELS OF ALS

3.3.1 Genetic Models

Genetic rodent models have been crucial in uncovering the pathways underlying ALS etiology and pathogenesis while also aiding in the identification of potential therapeutic targets. The number of available mouse models continues to expand, with each new mutation offering a unique pathogenetic perspective.

3.3.1.1 Superoxide Dismutase 1 (SOD1)-Based Rodent Models of ALS

One of the most extensively studied models of ALS is the SOD1G93A mouse, developed following the discovery of the first ALS-linked *SOD1* mutation in the 1990s. This model has been instrumental in preclinical research, significantly advancing our understanding of ALS pathogenesis, particularly its non-cell-autonomous nature. The SOD1G93A mouse exhibits key features observed in human ALS, including rapid motor neuron degeneration, paralysis, and death within four to five months. Notably, sex differences in disease progression have been reported, suggesting potential hormonal or genetic modulators of disease expression.

Despite its widespread use, this model has several limitations that must be considered in experimental design. Spontaneous loss of mutant SOD1 copies over time can lead to variability in disease onset and severity. Additionally, genetic drift and background strain differences may influence phenotypic outcomes, including pathology progression and lifespan.

To address these challenges, alternative SOD1 mouse models have been developed, including SOD1G37R, SOD1G86R, and SOD1G85R. These models differ in mutant SOD1 expression levels, disease onset, clinical manifestations, and lifespan. However, they share common pathological characteristics, such as the following (5–7):

- Early onset of symptoms
- Microgliosis and neuroinflammation
- Glutamate excitotoxicity

- Mitochondrial dysfunction
- Axonal transport defects
- Motor neuron loss leading to progressive paralysis and death

In addition to mouse models, SOD1-mutant rat models have also been developed, exhibiting neuropathological features similar to those observed in mice. As with mice, disease severity in rats correlates with mutant SOD1 transgene copies expressed. Among these, the G93A mutation induces a more aggressive phenotype compared to mutations such as H46R. Rats offer distinct advantages for preclinical research, including their larger body size, which facilitates surgical procedures, drug administration, and repeated sampling. However, these benefits are tempered by higher maintenance costs and greater logistical demands (8).

Together, SOD1-based rodent models continue to serve as indispensable tools for elucidating ALS pathomechanisms and for the preclinical evaluation of emerging therapeutic strategies.

3.3.1.2 TAR DNA-Binding Protein 43 (TDP-43)-Based Rodent Models of ALS

Another major advancement in ALS research has been the development of rodent models carrying mutations in TAR DNA-binding protein 43 (TDP-43). TDP-43 mutations account for approximately 4% of familial ALS cases and are also observed in a subset of sporadic cases. Notably, mutant TDP-43 is the principal component of ubiquitinated cytoplasmic inclusions found in the majority of ALS patients, making these models highly valuable for investigating neurotoxic mechanisms associated with ALS pathology (9).

Since the identification of TARDBP mutations in familial ALS, nearly 20 distinct mouse models have been generated. However, the development of these models has proven technically challenging due to the dose-sensitive nature of TDP-43 expression.

Complete TARDBP silencing results in neurodegeneration, while even minimal overexpression of wild-type or mutant TDP-43 induces widespread RNA dysregulation and cellular toxicity. Interestingly, dietary factors have been reported to influenc e disease phenotypes in TDP-43 mutant models (10).

The first transgenic *TDP-43* mouse models overexpressed wild-type or mutant (A315T and M337V) *TDP-43* cDNAs under the prion protein (*Prnp*) promoter. These mice developed ubiquitinated protein aggregates, leading to severe neuronal abnormalities, cytoplasmic inclusions, and spinal cord pathology. The severity of the phenotype was positively correlated with mutant *TDP-43* expression levels, with high-expression mice dying within weeks of birth (11).

Subsequent modifications to the *Prnp* promoter produced less aggressive phenotypes. For example, overexpression of M337V, Q331K, and wild-type *TDP-43* under the Thy1 promoter led to motor impairment by three months, including tremors, hind limb coordination deficits, and muscle weakness. Among these, the Q331K mutation was the least severe, showing only electromyography impairments at ten months, with minimal motor neuron loss, non-progressive axonal degeneration, and reduced neuromuscular junction damage (12).

By contrast, transgenic mice expressing the A315T mutation developed TDP-43-positive cytoplasmic inclusions, but degeneration halted around 20 months and was non-lethal. However, co-expression of *Prnp-Q311K* and wild-type TDP-43 alleles led to a much more severe phenotype, characterized by motor neuron loss, *TDP-43* inclusions, gliosis, and death within eight to ten weeks.

To improve physiological relevance, BAC (bacterial artificial chromosome) transgenic models were developed, incorporating TARDBP sequences under endogenous promoters. BAC-transgenic mice expressing A315T, G348C, or wild-type TDP-43 demonstrated the following:

- Motor deficits
- Age-related neuroinflammation (onset around 42 weeks)
- Spatial learning impairments (by one year)
- Ubiquitinated *TDP-43* inclusions in the spinal cord (by ten months)
- Normal lifespan

In parallel, rat models overexpressing wild-type or M337V TDP-43 under human TARDBP promoters were established. Rats expressing M337V exhibited early motor impairments, followed by paralysis and death before sexual maturity. In contrast, rats expressing comparable levels of wild-type TDP-43 did not develop paralysis by 200 days, supporting the notion that mutant TDP-43 is intrinsically more toxic than the wild-type protein.

In conclusion, TDP-43 rodent models provide critical insights into ALS mechanisms, particularly *TDP-43* mislocalization, aggregation, and neurodegeneration. However, their high phenotypic variability, influenced by expression levels and genetic background, highlights the complexity of *TDP-43*-linked ALS and the need for refined models.

3.3.1.3 Fused in Sarcoma (FUS)-Based Rodent Models of ALS

Fused in Sarcoma (FUS) is an RNA- and DNA-binding protein involved in diverse cellular processes, including transcription regulation, RNA splicing, and RNA transport. Mutations in the FUS gene have been implicated in both familial and sporadic forms of ALS. These mutations often result in cytoplasmic mislocalization and aggregation of FUS protein, which disrupts normal cellular function and contributes to motor neuron degeneration.

To investigate the pathogenic mechanisms associated with FUS mutations, a variety of rodent models have been developed using different genetic strategies.

Mouse Models:

1. Homozygous FUS Mutant Mice

   These models exhibit high mortality rates, largely due to apoptotic death of motor neurons and cytoplasmic FUS mislocalization.
2. Heterozygous *FUS* Mutant Mice

   Heterozygous models demonstrate progressive motor neuron loss without overt paralysis or early lethality. Despite the milder phenotype, they exhibit increased cytoplasmic FUS accumulation, offering a more gradual disease course that may better mimic certain clinical presentations of FUS-related ALS.
3. Transgenic Overexpression Models

   Transgenic mice have been generated using promoters such as microtubule-associated protein tau (MAPT) and Prnp to drive overexpression of either wild-type or mutant FUS. These models present with a range of pathological features, including the following:
   - Motor dysfunction
   - Protein aggregation
   - Neuroinflammatory responses

   Phenotypic severity varies depending on the promoter used, the level of transgene expression, and the specific FUS variant introduced.
4. Humanized Knock-In Models

   More physiologically relevant models incorporate patient-specific mutations, including those affecting 3′ splice site regulation. These knock-in models develop progressive motor dysfunction, muscle denervation, and reduced lifespan, offering important insights into mutation-specific disease mechanisms.

   Rat Models:

   AAV9-Mediated FUS Overexpression

   Transgenic rat models have also been developed using adeno-associated virus (AAV9) vectors to deliver and overexpress human FUS variants. These models exhibit the following:
   - Progressive motor impairment and respiratory dysfunction
   - Motor axon degeneration, neuronal loss, and gliosis
   - Cytoplasmic FUS aggregation, particularly pronounced in models expressing the R512C mutant, which leads to early-onset pathology

Overall, rodent models carrying *FUS* mutations have provided critical insights into the molecular and cellular mechanisms underlying *FUS*-related ALS. These models not only replicate key pathological features, such as cytoplasmic *FUS* mislocalization, neuroinflammation, and motor neuron degeneration but also serve as valuable tools for evaluating potential therapeutic strategies for ALS.

3.3.1.4 Chromosome 9 Open Reading Frame 72 (C9orf72)-Based Rodent Models of ALS

In addition to SOD1 and RNA-binding protein mutations, genetic studies have identified the C9orf72 gene on chromosome 9p21 as a major site of expansion in patients with ALS and FTD. The expanded hexanucleotide repeat (GGGGCC; G_4C_2) in C9orf72 is the most common genetic cause of familial ALS and FTD, collectively referred to as C9ALS-FTD. Although the full biological function of the C9orf72 protein is still being investigated, it has been implicated in autophagy regulation, endo-lysosomal trafficking, and immune homeostasis.

1. Early Transgenic Mouse Models

   The first C9orf72 repeat expansion mouse model was developed by Xu et al. in 2013. These mice demonstrated a broad spectrum of pathological features, including the following:
   - Neuromuscular junction damage
   - Hippocampal disorganization and cognitive impairments
   - Apoptosis and gait abnormalities
   - Protein inclusions of TDP-43 and dipeptide repeat proteins
   - Loss of Purkinje and cortical neurons
   - Astrogliosis, weight loss, hyperactivity, and motor deficits
2. Knockout Models to Investigate Loss-of-Function (LoF) Mechanisms

   To explore the hypothesis that C9orf72 loss contributes to neurodegeneration, several knockout (KO) models were generated, including the following:

   a. Neuron-Specific Knockout Models
      - Mice with selective neural loss of *C9orf72* were generated to assess neurodegeneration.
      - Surprisingly, these mice did not exhibit motor neuron degeneration, motor deficits, or decreased survival, suggesting that *C9orf72* loss alone may not be sufficient to drive ALS pathology.

   b. Constitutive Knockout Models
      - Generated using the Knock-Out Mouse Program (KOMP) and zinc finger nuclease (ZFN) technology.
      - These mice exhibit splenomegaly, lymphadenopathy, and age-related neuroinflammation, partially mirroring the immunopathology observed in patients with C9ALS.

   c. Immune Phenotype Knockout Models

      A third C9orf72 KO model developed a strong immune phenotype, characterized by the following:
      - Myeloid expansion and T-cell activation
      - Increased plasma cells and autoantibody production
      - Immune-mediated glomerulonephritis

   Although this model does not replicate classical ALS features, it suggests that C9orf72 loss may predispose to autoimmunity and systemic immune dysregulation.
3. BAC Transgenic Models of C9orf72 Repeat Expansion:

   To more accurately mimic human C9ALS, four BAC (bacterial artificial chromosome) transgenic mouse models expressing human C9orf72 repeat expansions were developed. These models shared several pathological features, including the following:
   - Nuclear RNA foci
   - Cytoplasmic repeat-associated non-ATG (RAN) translation inclusions

One model, developed by Liu et al. (2016), demonstrated a robust ALS-like phenotype with:

- Widespread TDP-43 inclusions (nuclear and cytoplasmic)
- Motor neuron loss and muscle denervation
- Hind limb paralysis and reduced survival

4. Rat Models of C9orf72 Mutation:
   a. C9orf72 Knockout (KO) Rats

      Using CRISPR/Cas9, C9orf72 KO rats were generated. Interestingly, these rats showed only minor alterations in cervical lymph nodes and spleen, with no motor neuron loss. However, when treated with low doses of kainic acid (KA) for four weeks, they developed the following:
      - Significant motor neuron loss
      - Progressive motor deficits

      This suggests that *C9orf72* loss may act as a sensitizing factor, making motor neurons more vulnerable to secondary insults such as excitotoxicity.

   b. C9orf72 Knock-In (KI) Rats

      A KI rat model was created by inserting 80 G_4C_2 repeats with human flanking sequences into the endogenous C9orf72 locus. This model displayed the following:
      - Reduced C9orf72 protein expression in multiple CNS regions
      - Onset of motor deficits at four months
      - Motor neuron loss leading to progressive neuromuscular dysfunction

This model provides further insight into LoF contributions to *C9orf72*-related neurotoxicity.

In conclusion, C9orf72 rodent models have become essential tools in preclinical ALS and FTD research. They have provided critical insights into disease mechanisms, including RNA toxicity, TDP-43 pathology, and immune dysregulation. The diversity of these models highlights the heterogeneous nature of C9orf72-associated ALS, underscoring the importance of careful selection based on the specific experimental objectives and phenotypic features under investigation.

3.3.1.5 ALS2-Based Rodent Models of ALS

Mutations in the ALS2 gene, which encodes the 184-kDa protein alsin, have been linked to autosomal recessive forms of ALS. Alsin is a ubiquitously expressed protein that contains three guanine-nucleotide-exchange factor (GEF)-like domains, and it plays an important role in intracellular signaling and vesicle trafficking.

Several pathogenic mutations in ALS2 have been identified, particularly within the regulator of chromatin condensation 1 (RCC1) domain, which is essential for the normal function of alsin. Among these, two frameshift mutations—261delA and 553delA—are associated with a juvenile-onset form of ALS, classified as ALS type 2 (ALS2). To investigate the role of alsin deficiency in ALS pathogenesis, researchers have developed ALS2-KO mouse models. However, these models have largely failed to replicate the hallmark features of motor neuron disease. The absence of significant motor neuron degeneration in these mice suggests that alsin deficiency alone may not be sufficient to trigger ALS pathology, indicating a potential need for additional genetic or environmental factors to drive neurodegeneration in ALS2-related cases (5).

3.3.1.6 Senataxin (SETX)-Based Rodent Models of ALS

Mutations in the senataxin (SETX) gene have been associated with a juvenile form of ALS, known as ALS type 4 (ALS4). Unlike many other ALS-linked proteins, senataxin is an RNA-binding protein without a low-complexity domain, making it structurally and functionally distinct. The SETX mutation mouse model, introduced in 2018 by Craig and collaborators, has provided new insights into the molecular mechanisms of this disease.

Two distinct SETX mouse models have been developed:

1. R2136H SETX Mutation (Transgenic Model)
2. L389S SETX Mutation (Knock-in Model)

Both models exhibit neuromuscular deficits, motor neuron degeneration, and TDP-43 cytoplasmic mislocalization and clearance, closely mirroring the pathology observed in human ALS4 patients.

Key Findings from *SETX* ALS4 Models:

- *Nuclear Membrane Abnormalities*: Motor neurons in SETX ALS4 mice display structural alterations in the nuclear membrane, linked to disruptions in nucleocytoplasmic transport proteins Ran and RanGAP1.
- *Impaired Nucleocytoplasmic Trafficking*: Studies on SETX ALS4 cortical neurons reveal delayed nuclear import, suggesting that impaired transport between the nucleus and cytoplasm may contribute to motor neuron degeneration (13).

These results demonstrate that SETX ALS4 mice recapitulate ALS disease phenotypes, including TDP-43 mislocalization. The link between SETX dysfunction and common pathways of ALS motor neuron degeneration offers valuable insights for future research.

3.3.1.7 Optineurin (OPTN)-Based Rodent Models of ALS

Mutations in the optineurin (OPTN) gene have been identified in both familial and sporadic cases of ALS. OPTN is a multifunctional protein that plays a pivotal role in selective autophagy, facilitating the clearance of damaged proteins and organelles, an essential process for maintaining neuronal health and homeostasis. A number of point mutations in the OPTN gene have been associated with ALS pathogenesis. Interestingly, some of these variants are also implicated in primary open-angle glaucoma. Notable ALS-associated OPTN mutations include the following:

- Homozygous deletion of exon 5
- Missense mutation R96L
- Homozygous Q398X nonsense mutation
- Heterozygous E478G missense mutation

To explore the functional impact of OPTN mutations in vivo, researchers developed a KI mouse model in which the endogenous OPTN gene was replaced by the $OPTN^{D477N}$ mutant allele. This particular mutation is considered equivalent to a non-pathogenic variant in humans. Notably, these mice did not display motor deficits or neurodegeneration, suggesting that not all OPTN mutations contribute equally to disease progression (14,15).

3.3.1.8 Ubiquilin-2 (UBQLN2)- Based Rodent Models of ALS

Mutations in the ubiquilin-2 (UBQLN2) gene have been associated with both ALS and FTD. Several transgenic mouse models have been developed to study the pathogenic mechanisms of these mutations, revealing a range of cognitive and neurodegenerative phenotypes. Mouse models carrying *UBQLN2* mutations exhibit cognitive deficits, dendritic spinopathy, and ubiquilin-2 inclusion bodies in the hippocampus. However, these models do not develop *TDP-43* pathology or motor neuron loss.

1. Knock-In and Mutant-Specific Models

   KI mouse models of UBQLN2 recapitulate the cognitive deficits observed in human disease but generally lack motor symptoms. In contrast, mice expressing specific mutations, such as hUBQLN2P497S or hUBQLN2P506T, show both motor neuron loss and cognitive impairments, suggesting a mutation-dependent variation in disease presentation.

2. Double Transgenic UBQLN2 and TDP-43 Model

   A double transgenic mouse model harboring both hUBQLN2P497H and hTDP-43G348C mutations has been developed to investigate potential interactions between *UBQLN2* and *TDP-43* pathology. This model exhibits motor neuron loss, muscle atrophy, and progressive motor and cognitive impairments during aging.

3. Rat Models of UBQLN2 Mutations

   Complementary rat models carrying mutant UBQLN2 alleles have also been established. These animals display the following:
   - Cognitive deficits
   - Hippocampal ubiquilin-2 aggregate formation
   - Neuronal loss (16,17)

3.3.1.9 Profilin 1 (PFN1)-Based Rodent Models of ALS

Mutations in the profilin 1 (PFN1) gene, which encodes a key regulator of actin filament dynamics and cytoskeletal organization, have been linked to ALS. However, the precise mechanisms by which these mutations contribute to the disease remain unclear. Both LoF and gain-of-function (GoF) mechanisms have been proposed.

1. PFN1G118V Mouse Model

   One of the earliest ALS mouse models involving PFN1 utilized hemizygous transgenic mice expressing the PFN1G118V mutation. These mice exhibited several key pathological features consistent with ALS, including the following:
   - Motor neuron degeneration and loss
   - Intracellular aggregation of mutant PFN1
   - Glial cell activation
   - Muscle atrophy and weight loss
   - Reduced survival
   - Motor dysfunction becomes apparent around 120–130 days of age, with the end-stage of the disease typically occurring around 165–210 days
2. PFN1C71G Mouse Models

   The PFN1C71G mutation has been studied using multiple transgenic approaches:
   - Standard Transgenic Line
     - Hemizygous mice exhibited mild motor weakness by approximately 350 days.
     - Homozygous mice exhibited a more aggressive disease course, with symptoms appearing at 150 days and full paralysis by 320 days
   - Prnp-Driven and Thy1-PFN1C71G Crossed Model

   Crossing a Prnp-driven PFN1 transgenic line with a Thy1-PFN1C71G homozygous line-accelerated ALS pathology.
   - Hb9-PFN1C71G Model

   A novel Hb9-driven PFN1C71G mouse model was developed to specifically target α-motor neurons in the spinal cord during development. Adult mice in this model developed significant motor deficits and demonstrated a marked reduction in spinal motor neuron populations, underscoring the critical role of PFN1 in neurodevelopment (18–20).

3.3.1.10 Valosin-Containing Protein (VCP)-Based Rodent Models of ALS

Mutations in the valosin-containing protein (VCP) gene are associated with several neurodegenerative disorders, including ALS. To study VCP-related pathologies, researchers have developed transgenic mouse models carrying VCP mutations, which replicate key features of human disease.

1. VCPR155H Heterozygous Mouse Model

   The VCPR155H heterozygous mouse exhibits age-dependent motor neuron degeneration in the ventral horn, accompanied by the following:
   - TDP-43-positive inclusions
   - Mitochondrial dysfunction
   - Astrocyte activation in muscles, bones, and brain

   Although this model does not develop **a** rapidly progressive or fatal ALS-like pathology, it recapitulates critical aspects of the disease, making it a valuable tool for studying early-stage neurodegeneration.
2. VCPR155H Homozygous Mouse Model

   The homozygous VCPR155H model displays similar pathological alterations as the heterozygous model but with earlier onset and greater severity. These mice typically exhibit the following:
   - Earlier onset and more severe neurodegeneration
   - Pronounced skeletal muscle atrophy and bone abnormalities
   - Drastically reduced lifespan, typically surviving less than one month (21–24)

3.3.1.11 Vesicle-Associated Membrane Protein-Associated Protein B-Based Rodent Models of ALS

Mutations in the Vesicle-Associated Membrane Protein-associated protein B (VAPB) gene have been identified in several ALS patients, making VAPB mutant rodent models valuable for studying disease mechanisms. VAPB is a multifunctional protein involved in processes such as membrane trafficking, neurotransmitter release, lipid transport, and unfolded protein response (UPR). Several ALS-associated *VAPB* mutations have been identified, including P56S, T46I, del160, D130E, A145V, S160Δ, and V234I. Among these, the P56S mutation is the most extensively studied and well-characterized.

3.3.1.11.1 Transgenic and Knock-In VAPB Mouse Models

To explore the pathophysiological effects of VAPB mutations, transgenic mice overexpressing either wild-type or P56S-mutant human VAPB have been developed using both neuron-specific and ubiquitous promoters. Notably, only the mice expressing P56S-VAPB developed motor deficits, highlighting a mutation-specific pathogenic effect.

To better replicate human pathology, KO P56S-VAPB mouse models were later developed by Larroquette et al. These mice displayed the following:

- Late-onset motor behavior impairments, resembling symptoms seen in human ALS cases
- Ubiquitinated protein accumulation
- ER stress
- Autophagic dysfunction (24,25)

3.3.2 Non-Genetic Models

3.3.2.1 Beta-N-methylamino-L-alanine (L-BMAA)-Induced ALS Rodent Models

While genetic mutations play a significant role in ALS pathogenesis, environmental factors have also been implicated in disease onset and progression. One such environmental risk factor is exposure to beta-N-methylamino-L-alanine (L-BMAA), a non-proteinogenic amino acid produced by cyanobacteria and found in cycad seeds. L-BMAA exposure has been strongly associated with ALS-parkinsonism-dementia complex (ALS-PDC), a condition endemic to the Chamorro population of Guam Island, commonly referred to as Guam disease.

To investigate the neurotoxic effects of L-BMAA and its potential role in ALS, researchers have developed several rodent models that mimic key pathological features of the disease. These models have provided valuable insights into disease mechanisms and potential therapeutic targets.

One experimental approach involves administering L-BMAA to young rats to induce ALS-like phenotypes. In a study utilizing this model, 21-day-old rats received intraperitoneal injections of L-BMAA at varying doses (up to 350 mg/kg/day) for five consecutive days. Neurological and functional assessments were conducted over eight months. At doses exceeding 200 mg/kg, rats exhibited pronounced motor deficits, including reduced ambulation, impaired tail lift, diminished grip strength, and poor performance on the rotarod test. Gait abnormalities were also evident through altered footprint analysis.

Histological examinations revealed severe neurodegenerative changes, including motor neuron , the accumulation of high molecular weight TDP-43 in the lumbar spinal cord and motor cortex, and reductions in both motor cortex and muscle volume. Additionally, structural alterations were observed in the bulbar nuclei, further supporting the relevance of this model to ALS pathology. Another approach to modeling L-BMAA-induced neurotoxicity involves chronic administration via intrathecal infusion. In this model, adult rats received continuous intrathecal infusion of L-BMAA (5 mM) for 30 days using mini-osmotic pumps. This prolonged exposure resulted in substantial degeneration of motor neurons in the ventral horn of the spinal cord, accompanied by cytosolic mislocalization of TDP-43, reductions in the glutamate transporter GLT-1, and marked reactive gliosis. Microglial infiltration and astrogliosis were also observed, indicating an inflammatory response (26–29).

REFERENCES

1. Kiernan MC, Vucic S, Cheah BC, Turner MR, Eisen A, Hardiman O, et al. Amyotrophic lateral sclerosis. Lancet. 2011 Mar;377(9769):942–55.
2. Al-Chalabi A, van den Berg LH, Veldink J. Gene discovery in amyotrophic lateral sclerosis: implications for clinical management. Nat Rev Neurol. 2017 Feb;13(2):96–104.
3. Swarup V, Julien JP. ALS pathogenesis: recent insights from genetics and mouse models. Prog Neuropsychopharmacol Biol Psychiatry. 2011 Mar;35(2):363–9.
4. Morrice JR, Gregory-Evans CY, Shaw CA. Animal models of amyotrophic lateral sclerosis: a comparison of model validity. Neural Regen Res. 2018 Dec;13(12):2050–4.
5. Hadano S, Otomo A, Kunita R, Suzuki-Utsunomiya K, Akatsuka A, Koike M, et al. Loss of ALS2/Alsin exacerbates motor dysfunction in a SOD1-expressing mouse ALS model by disturbing endolysosomal trafficking. PLoS One. 2010 Mar;5(3):e9805.
6. Wang XX, Chen WZ, Li C, Xu RS. Current potential pathogenic mechanisms of copper-zinc superoxide dismutase 1 (SOD1) in amyotrophic lateral sclerosis. Rev Neurosci. 2024 Jul;35(5):549–63.
7. Furukawa Y, Tokuda E. Does wild-type Cu/Zn-superoxide dismutase have pathogenic roles in amyotrophic lateral sclerosis? Transl Neurodegener. 2020 Aug;9(1):33.
8. Zhou L, Xie M, Wang X, Xu R. The usage and advantages of several common amyotrophic lateral sclerosis animal models. Front Neurosci. 2024;18:1341109.
9. Suk TR, Rousseaux MWC. The role of TDP-43 mislocalization in amyotrophic lateral sclerosis. Mol Neurodegener. 2020 Aug;15(1):45.
10. Sreedharan J, Blair IP, Tripathi VB, Hu X, Vance C, Rogelj B, et al. TDP-43 mutations in familial and sporadic amyotrophic lateral sclerosis. Science. 2008 Mar;319(5870):1668–72.
11. Xu YF, Zhang YJ, Lin WL, Cao X, Stetler C, Dickson DW, et al. Expression of mutant TDP-43 induces neuronal dysfunction in transgenic mice. Mol Neurodegener. 2011 Oct;6:73.
12. Bonifacino T, Zerbo RA, Balbi M, Torazza C, Frumento G, Fedele E, et al. Nearly 30 years of animal models to study amyotrophic lateral sclerosis: a historical overview and future perspectives. Int J Mol Sci. 2021 Nov;22(22).
13. Tsui A, Kouznetsova VL, Kesari S, Fiala M, Tsigelny IF. Role of senataxin in amyotrophic lateral sclerosis. J Mol Neurosci. 2023 Dec;73(11–12):996–1009.

14. Prtenjaca N, Rob M, Alam MS, Markovinovic A, Stuani C, Buratti E, et al. Optineurin deficiency and insufficiency lead to higher microglial TDP-43 protein levels. Int J Mol Sci. 2022 Jun;23(12).
15. Zhao S, Chen R, Gao Y, Lu Y, Bai X, Zhang J. Fundamental roles of the Optineurin gene in the molecular pathology of amyotrophic lateral sclerosis. Front Neurosci. 2023;17:1319706.
16. Whiteley AM, Prado MA, de Poot SAH, Paulo JA, Ashton M, Dominguez S, et al. Global proteomics of Ubqln2-based murine models of ALS. J Biol Chem. 2021;296:100153.
17. Renaud L, Picher-Martel V, Codron P, Julien JP. Key role of UBQLN2 in pathogenesis of amyotrophic lateral sclerosis and frontotemporal dementia. Acta Neuropathol Commun. 2019 Jul;7(1):103.
18. Brettle M, Stefen H, Djordjevic A, Fok SYY, Chan JW, van Hummel A, et al. Developmental expression of mutant PFN1 in motor neurons impacts neuronal growth and motor performance of young and adult mice. Front Mol Neurosci. 2019;12:231.
19. Kiaei M, Balasubramaniam M, Govind Kumar V, Shmookler Reis RJ, Moradi M, Varughese KI. ALS-causing mutations in profilin-1 alter its conformational dynamics: a computational approach to explain propensity for aggregation. Sci Rep. 2018 Aug;8(1):13102.
20. Fil D, DeLoach A, Yadav S, Alkam D, MacNicol M, Singh A, et al. Mutant Profilin1 transgenic mice recapitulate cardinal features of motor neuron disease. Hum Mol Genet. 2017 Feb;26(4):686–701.
21. Chu S, Xie X, Payan C, Stochaj U. Valosin containing protein (VCP): initiator, modifier, and potential drug target for neurodegenerative diseases. Mol Neurodegener. 2023 Aug;18(1):52.
22. Nalbandian A, Donkervoort S, Dec E, Badadani M, Katheria V, Rana P, et al. The multiple faces of valosin-containing protein-associated diseases: inclusion body myopathy with Paget's disease of bone, frontotemporal dementia, and amyotrophic lateral sclerosis. J Mol Neurosci. 2011 Nov;45(3):522–31.
23. Yin HZ, Nalbandian A, Hsu CI, Li S, Llewellyn KJ, Mozaffar T, et al. Slow development of ALS-like spinal cord pathology in mutant valosin-containing protein gene knock-in mice. Cell Death Dis. 2012 Aug;3(8):e374.
24. Kamemura K, Kozono R, Tando M, Okumura M, Koga D, Kusumi S, et al. Secretion of endoplasmic reticulum protein VAPB/ALS8 requires topological inversion. Nat Commun. 2024 Oct;15(1):8777.
25. Vucic S, Kiernan MC. Pathophysiology of neurodegeneration in familial amyotrophic lateral sclerosis. Curr Mol Med. 2009 Apr;9(3):255–72.
26. Karamyan VT, Speth RC. Animal models of BMAA neurotoxicity: a critical review. Life Sci. 2008 Jan 30;82(5–6):233–46.
27. Liu XQ, Rush T, Zapata J, Lobner D. β-N-methylamino-l-alanine induces oxidative stress and glutamate release through action on system Xc-. Exp Neurol. 2009 Jun;217(2):429–33.
28. Zhou X, Escala W, Papapetropoulos S, Grace Zhai R. β-N-methylamino-L-alanine induces neurological deficits and shortened life span in Drosophila. Toxins (Basel). 2010 Nov;2(11):2663–79.
29. Lopicic S, Svirčev Z, Palanački Malešević T, Kopitović A, Ivanovska A, Meriluoto J. Environmental neurotoxin β-N-methylamino-L-alanine (BMAA) as a widely occurring putative pathogenic factor in neurodegenerative diseases. Microorganisms. 2022 Dec 1;10(12).

4 Rodent Models of Frontotemporal Dementia

Fatemeh Mokhtari Naseri

4.1 INTRODUCTION

Frontotemporal lobar degeneration (FTLD) leads to frontotemporal dementia (FTD), which is the second most prevalent type of primary degenerative dementia following Alzheimer's disease. FTD mainly affects individuals in their midlife, typically between 45 and 65 years of age. However, cases have been reported in both younger individuals and the elderly. FTD accounts for approximately 20% of presenile dementia cases and affects both genders equally. The disease progression is highly variable, with an average duration of around eight years, though it can range from 2 to 20 years.

A familial history of dementia is observed in nearly half of the affected individuals, highlighting a strong genetic component in disease pathogenesis. Characterized by profound behavioral and cognitive impairments, FTD primarily manifests through significant alterations in personality, social behavior, and communication. Unlike Alzheimer's disease, memory functions are relatively preserved in the early stages, while executive dysfunction, disinhibition, and language deficits become prominent features. The disorder encompasses a spectrum of pathological subtypes, with behavioral variant FTD (bvFTD) being the most prevalent, often associated with TDP-43 proteinopathy. Genetic studies have identified key mutations in genes such as MAPT, GRN, and chromosome 9 open reading frame 72 (C9orf72), reinforcing the importance of genetic factors in disease progression and pathology (1–3).

Rodent models have become indispensable tools for studying FTD, providing valuable insights into its molecular mechanisms and potential therapeutic targets. Mouse models, in particular, are widely utilized due to their genetic manipulability and neuroanatomical similarities to humans. Studies focusing on the salience network, which is disrupted in FTD, along with behavioral assessments targeting social dysfunction, repetitive behaviors, and emotional regulation, have been instrumental in bridging the gap between preclinical findings and human pathology. Despite inherent limitations, such as shorter lifespans and species-specific differences in cognition and behavior, rodent models remain crucial for understanding FTD pathology and evaluating novel treatment strategies. As research continues to refine these models and uncover new genetic contributors, they hold promise for advancing therapeutic approaches and improving outcomes for individuals affected by FTD (4,5).

This chapter provides an overview of FTD, followed by a discussion of various rodent models used to study the disease and their relevance in advancing our understanding of its pathophysiology and treatment approaches.

4.2 OVERVIEW OF FTD

FTD is a neurodegenerative disorder characterized by progressive changes in behavior, personality, and language, while memory and perceptual functions are relatively preserved during the early stages. The disorder encompasses a spectrum of clinical, pathological, and genetic variants, making it one of the most heterogeneous forms of dementia.

A. Core Clinical Features
    a. Behavioral Alterations

DOI: 10.1201/9781032676524-5

While perceptual and memory functions remain intact early on, patients show distinct behavioral abnormalities:

- *Decline in Social Behavior*: Patients may disregard social norms, engage in inappropriate behaviors, and show tactlessness.
- *Impaired Regulation of Personal Conduct*: Behavior may oscillate between passivity and hyperactivity, sometimes leading to excessive wandering or pacing.
- *Emotional Blunting*: There is a diminished ability to express emotions, including both basic emotions (joy, sadness) and complex social emotions (empathy, embarrassment).
- Lack of Insight: Patients often fail to recognize one's symptoms and exhibit inappropriate emotional reactions.

b. Changes in Eating Habits

Atypical eating behaviors are common, including the following:

- Hyperphagia or compulsive eating
- Intense cravings, particularly for sweet foods

c. Repetitive or Stereotyped Behaviors

Patients may develop the following:

- *Motor stereotypes*: Humming, hand-rubbing, foot-tapping
- *Utilization behaviors*: Inappropriate or out-of-context use of objects

d. Communication and Cognitive Impairments

Patients may develop the following:

- *Reduced Speech Output*: Progressive reduction in spontaneous speech, mutism, echolalia, and use of stereotyped phrases
- *Executive Dysfunction*: Impaired attention, abstraction, and cognitive flexibility, despite early preservation of memory and visuospatial skills

B. Pathological and Genetic Insights

Advances in FTD research have led to the identification of disease-associated proteins and genetic mutations. Historically, tau protein was the first major pathological marker, particularly in familial FTD. However, since 2004, other key proteins and mutations have been recognized:

Major pathological proteins:

- TDP-43 (TAR DNA-binding protein 43)
- Tau
- FUS (Fused in Sarcoma)

Key genetic mutations associated with familial FTD:

- MAPT (microtubule-associated protein tau)
- GRN (progranulin)
- C9orf72
- VCP (valosin-containing protein)
- FUS
- CHMP2B
- TARDBP

C. Clinical Subtypes of FTD

FTD encompasses a range of clinical phenotypes, each with distinct symptoms and pathological correlates:

- *Behavioral Variant FTD (bvFTD)*: The most prevalent form, primarily affecting behavior and personality.
- *Primary Progressive Aphasia (PPA)*: Characterized by language impairment, with two subtypes:
  - Semantic variant (loss of word meaning).
  - Nonfluent/agrammatic variant (difficulty producing grammatically correct speech).

- *Progressive Supranuclear Palsy (PSP)*: Marked by movement disturbances and gaze palsy.
- *Corticobasal Syndrome (CBS)*: A syndrome with asymmetric movement disorders and cognitive impairment.
- *FTD with Amyotrophic Lateral Sclerosis (FTD-ALS)*: A combination of behavioral dysfunction and motor neuron disease.

Among these, bvFTD is the most studied due to its diverse pathological mechanisms. The predominant pathology is FTLD with TDP-43 pathology (FTLD-TDP), with further subtypes classified based on the cellular distribution of TDP-43. Other variants include FTLD-tau, FTLD-FUS, and rarer forms with ubiquitin-positive inclusions unrelated to tau, TDP-43, or FUS (6–8).

4.3 RODENT MODELS OF FTD

4.3.1 TDP-Linked FTD

4.3.1.1 GRN-Based Rodent Models of FTD

Mutations within the GRN gene, which encodes the progranulin protein, account for around 5% of FTD (including some sporadic instances) and are one of the primary genetic contributors to FTLD-TDP. FTD linked to GRN mutations typically presents several years earlier than other cases of FTLD-TDP and is less commonly associated with ALS. Pathologically, GRN mutations predominantly exhibit type A FTLD-TDP, recognized as the most prevalent FTLD-TDP subtype. As a result, progranulin-focused mouse models are particularly relevant for a wide range of FTD cases.

To date, approximately 70 distinct mutations in *GRN* have been identified, all leading to reduced progranulin levels or loss of its functional capacity. The majority of these mutations, including the frequently observed R493X mutation, lead to the production of unstable mRNA that is subjected to degradation via nonsense-mediated decay. Consequently, this process results in haploinsufficiency, meaning that a single functional copy of the gene is insufficient for normal progranulin production. Some mutations in the GRN gene lead to a complete loss of the gene's function, while others result in mutant progranulin proteins that either cannot be secreted from the cell or undergo improper cleavage. Due to this loss-of-function characteristic, progranulin-deficient mice are valuable models for studying the associated disorders.

Accordingly, several lines of progranulin knockout (*GRN–/–*) mice have been developed, all of which exhibit early-onset social behavior abnormalities without significant impairments in general health or motor function. However, hippocampus-dependent learning and memory functions, generally remain intact until later stages of life.

Early studies investigating the neuropathological aspects of *GRN–/–* mice highlight potential roles for both protein mishandling and neuroinflammation in this model.

- *Protein Mishandling*: This includes the accumulation of ubiquitinated proteins, lipofuscin, and, in some cases, phosphorylated TDP-43.
- *Neuroinflammation*: GRN–/– mice exhibit pronounced astrogliosis and microgliosis across all reported lines. Additionally, macrophages derived from these mice show heightened activity upon stimulation, and the mice themselves demonstrate increased susceptibility to inflammatory arthritis mediated by tumor necrosis factor-α (TNF-α) (9–12).

4.3.1.2 VCP-Based Rodent Models of FTD

Mutations in the VCP gene, which encodes valosin-containing protein (VCP) in humans; p97 in mice, are implicated in a subset of FTD cases. While these mutations can result in pure FTD, they more commonly cause a multisystem proteinopathy known as inclusion body myopathy with Paget disease of bone and FTD (IBMPFD). VCP/p97 is a ubiquitously expressed ATPase belonging to the ATPases associated with diverse cellular activities (AAA+) family. It plays an essential role in:

- Protein quality control, especially through the ubiquitin-proteasome system (UPS)
- Endoplasmic reticulum-associated degradation (ERAD)
- Autophagy

Given its broad involvement in cellular homeostasis, complete loss of VCP function is embryonically lethal, underscoring its vital role in development and cell survival.

FTD associated with VCP mutations typically presents in the mid-to-late 1950s, with variable phenotypic expression that includes the following:

- bvFTD
- Semantic variant PPA (svPPA)

FTD caused by VCP mutations is neuropathologically defined by FTLD with TDP-43 pathology (FTLD-TDP), specifically type D, which is unique to VCP-related cases and not seen in sporadic FTD or FTD caused by other genetic mutations. FTLD-TDP type D is characterized by:

- TDP-43-positive neuronal intranuclear inclusions
- Sparse neuronal cytoplasmic inclusions

This pattern of pathology sets it apart from the more common types observed in other forms of FTD.

A. VCP Mouse Models

To study VCP-related FTD and multisystem degeneration, researchers have developed mouse models harboring human-relevant mutations:

a. R155H Knock-In Mouse Model

A knock-in of the R155H mutation, one of the most prevalent disease-causing mutations in humans, was introduced into the endogenous mouse VCP gene. This model takes advantage of the high structural conservation between mouse and human VCP/p97. Although it shows molecular and histological hallmarks of disease, cognitive impairments, such as novel object recognition deficits, are not observed up to 15 months of age, suggesting a slowly progressive or subclinical disease course in this model (13–17).

4.3.1.3 TARDBP-Based Rodent Models of FTD

Mutations in the TARDBP gene, which encodes the protein TDP-43, are strongly associated with FTD. This connection is particularly evident in the subtype characterized by TDP-43 proteinopathy (FTLD-TDP), which encompasses most forms of ALS and accounts for almost half of all FTD cases.

Studies employing loss-of-function approaches have revealed the essential role of TDP-43 in development and homeostasis:

- Complete knockout (Tardbp–/–): Results in embryonic lethality, confirming that TDP-43 is indispensable for survival.
- Heterozygous knockout (Tardbp+/–): Maintains normal TDP-43 protein levels without significant neuropathology, though subtle age-dependent motor deficits may occur.
- Conditional knockout: Postnatal deletion (at four to six weeks of age) leads to mortality within nine days, primarily due to excessive fat oxidation and weight loss, highlighting TDP-43's role in metabolic regulation.

In addition to knockout models, various transgenic TDP-43 mouse models have been generated. Initial models employed heterologous promoters to drive systemic overexpression of human TDP-43 within the nervous system. These mice exhibited severe phenotypes, including profound motor deficits and premature death within weeks to months, largely due to high expression levels in the spinal cord.

Studies of pan-neuronal TDP-43 transgenic mice revealed that layer V of the frontal cortex was particularly vulnerable, displaying ubiquitinated neuronal inclusions, astrogliosis, and neuronal loss. Subsequent models utilized more selective promoters to restrict TDP-43 expression to the forebrain, thereby minimizing confounding motor phenotypes. Notably, even wild-type TDP-43 expression in the forebrain, driven by the CaMKII promoter, was sufficient to induce neurodegeneration and behavioral deficits. Specific neuronal populations displayed differential susceptibility; for instance, a transgenic line expressing wild-type human TDP-43 at 0.8 times endogenous levels exhibited over 75% loss of dentate granule neurons, whereas CA1 pyramidal neurons remained more resilient. It should be noted that transgenic mice expressing the entire human TDP-43 gene under its native promoter demonstrated a threefold increase in TDP-43 levels compared to endogenous expression. Both wild-type and mutant TDP-43 variants in these models led to the onset of astrocytosis by five months of age, followed by progressive cognitive and motor impairments beginning at seven months. By ten months, pathological hallmarks of FTLD-TDP, including ubiquitinated TDP-43 inclusions, were evident.

The phenotypic features observed in these TDP-43 mouse models provide valuable insights into the role of TDP-43 in FTD. Future research should focus on identifying which neuropathological features best correlate with neuronal degeneration and behavioral dysfunction. Although many models recapitulate key aspects of FTLD-TDP pathology, including nuclear and cytoplasmic TDP-43 inclusions, their correlation with functional deficits remains unclear, underscoring the need to explore additional underlying mechanisms. Additionally, a key challenge in TDP-43 modeling is the impact of transgenic human TDP-43 expression on endogenous mouse TDP-43 levels. The autoregulatory mechanisms governing TDP-43 expression complicate attempts to assess the functional consequences of transgenic overexpression, necessitating the development of more refined genetic models (18–23).

4.3.2 MAPT-Based Rodent Models of FTD

Tau is the first identified molecule linked to both the neuropathology and genetics of FTD. Cases characterized by tau pathology are classified as FTLD with tau (FTLD-tau). When these cases are associated with mutations in the microtubule-associated protein tau (MAPT) gene, they are more specifically referred to as FTD and parkinsonism linked to chromosome 17 (FTDP-17T). Mutations in the MAPT gene are among the three most prominent genetic causes of familial FTD, along with GRN and C9orf72. Individuals carrying MAPT mutations consistently exhibit tau accumulation and FTLD-tau pathology, often mirroring the clinical phenotypes observed in sporadic FTLD-tau cases. Given these extensive associations, mouse models of tau hold significant potential for studying FTD pathogenesis and therapeutic interventions (24–26).

4.3.3 C9orf72-Based Rodent Models of FTD

The expansion of a hexanucleotide repeat (GGGGCC, or G4C2) within an intron of the C9orf72 gene has been identified as a genetic factor associated with both amyotrophic lateral sclerosis (ALS) and FTD. To elucidate the molecular mechanisms underlying C9orf72-related neurodegeneration, several experimental models have been developed. One such model involved the knockout of the C9orf72 ortholog, 3110043O21Rik, in neurons and glial cells at embryonic day 10.5. This approach revealed no evident neurodegenerative changes, suggesting that the absence of C9orf72 alone is insufficient to precipitate the disease (27). This finding has led researchers to explore a potential gain-of-function mechanism associated with the C9orf72 expansion. Supporting this theory, studies have utilized adeno-associated virus (AAV) to overexpress a minigene consisting of 66 G4C2 repeats in mice. At six months of age, several pathological markers such as TDP-43 inclusions, loss of neurons in the cortex and cerebellum, astrogliosis, and a reduction in body weight. Additionally, the affected mice exhibited behavioral abnormalities, including hyperactivity, anxiety-like behaviors, and motor deficits (28). Evidence continues to accumulate indicating that the overexpression of C9orf72 RNAs or RAN proteins may contribute to gain-of-function effects. To further investigate these effects, a BAC transgenic mouse model containing the complete C9orf72 gene, along with significant flanking sequences, was created. This model facilitates the expression of both sense and antisense transcripts driven by human endogenous promoters. Notably, these symptoms included gait abnormalities, hindlimb weakness, and paralysis, accompanied by kyphosis and fluctuations in activity levels. Moreover, the mice exhibited hyperactivity when provoked, along with clasping behavior and occasional seizures. Neuropathological evaluations revealed significant neurodegenerative changes across several brain regions and the spinal cord. A marked loss of neurons was observed in the hippocampus, underscoring the severity of the condition. End-stage mice presented acute symptoms of neuroinflammation, along with TDP-43 inclusions characterized by nuclear and cytoplasmic aggregates in degenerating neurons throughout the brain (29).

REFERENCES

1. Bang J, Spina S, Miller BL. Frontotemporal dementia. Lancet. 2015 Oct;386(10004):1672–82.
2. Hogan DB, Jetté N, Fiest KM, Roberts JI, Pearson D, Smith EE, et al. The prevalence and incidence of frontotemporal dementia: a systematic review. Can J Neurol Sci. 2016 Apr;43 Suppl 1:S96–109.
3. Greaves C V, Rohrer JD. An update on genetic frontotemporal dementia. J Neurol. 2019 Aug;266 (8):2075–86.
4. Solomon DA, Mitchell JC, Salcher-Konrad MT, Vance CA, Mizielinska S. Review: Modelling the pathology and behaviour of frontotemporal dementia. Neuropathol Appl Neurobiol. 2019 Feb;45(1):58–80.
5. Nolan M, Talbot K, Ansorge O. Pathogenesis of FUS-associated ALS and FTD: insights from rodent models. Acta Neuropathol Commun. 2016 Sep;4(1):99.
6. Muñoz-Neira C, Tedde A, Coulthard E, Thai NJ, Pennington C. Neural correlates of altered insight in frontotemporal dementia: a systematic review. Neuroimage Clin. 2019;24:102066.
7. Barker MS, Gottesman RT, Manoochehri M, Chapman S, Appleby BS, Brushaber D, et al. Proposed research criteria for prodromal behavioural variant frontotemporal dementia. Brain. 2022 Apr;145(3):1079–97.
8. Lanata SC, Miller BL. The behavioural variant frontotemporal dementia (bvFTD) syndrome in psychiatry. J Neurol Neurosurg Psychiatry. 2016 May;87(5):501–11.
9. Piaceri I, Imperiale D, Ghidoni E, Atzori C, Bagnoli S, Ferrari C, et al. Novel GRN mutations in Alzheimer's disease and frontotemporal lobar degeneration. J Alzheimers Dis. 2018;62(4):1683–9.
10. Marsan E, Velmeshev D, Ramsey A, Patel RK, Zhang J, Koontz M, et al. Astroglial toxicity promotes synaptic degeneration in the thalamocortical circuit in frontotemporal dementia with GRN mutations. J Clin Invest. 2023 Mar;133(6).
11. Davis SE, Cook AK, Hall JA, Voskobiynyk Y, Carullo N V, Boyle NR, et al. Patients with sporadic FTLD exhibit similar increases in lysosomal proteins and storage material as patients with FTD due to GRN mutations. Acta Neuropathol Commun. 2023 Apr;11(1):70.
12. Huin V, Barbier M, Bottani A, Lobrinus JA, Clot F, Lamari F, et al. Homozygous GRN mutations: new phenotypes and new insights into pathological and molecular mechanisms. Brain. 2020 Jan;143(1):303–19.
13. Scarian E, Fiamingo G, Diamanti L, Palmieri I, Gagliardi S, Pansarasa O. The role of VCP mutations in the spectrum of amyotrophic lateral sclerosis-frontotemporal dementia. Front Neurol. 2022;13:841394.
14. Saracino D, Clot F, Camuzat A, Anquetil V, Hannequin D, Guyant-Maréchal L, et al. Novel VCP mutations expand the mutational spectrum of frontotemporal dementia. Neurobiol Aging. 2018 Dec;72:187.e11–e14.

15. Pfeffer G, Lee G, Pontifex CS, Fanganiello RD, Peck A, Weihl CC, et al. Multisystem proteinopathy due to VCP mutations: a review of clinical heterogeneity and genetic diagnosis. Genes (Basel). 2022 May;13(6).
16. Abrahao A, Abath Neto O, Kok F, Zanoteli E, Santos B, Pinto WBV de R, et al. One family, one gene and three phenotypes: a novel VCP (valosin-containing protein) mutation associated with myopathy with rimmed vacuoles, amyotrophic lateral sclerosis and frontotemporal dementia. J Neurol Sci. 2016 Sep;368:352–8.
17. Andrés-Benito P, Gelpi E, Povedano M, Santpere G, Ferrer I. Gene expression profile in frontal cortex in sporadic frontotemporal lobar degeneration-TDP. J Neuropathol Exp Neurol. 2018 Jul;77(7):608–27.
18. Bright F, Chan G, van Hummel A, Ittner LM, Ke YD. TDP-43 and inflammation: implications for amyotrophic lateral sclerosis and frontotemporal dementia. Int J Mol Sci. 2021 Jul;22(15).
19. de Boer EMJ, Orie VK, Williams T, Baker MR, De Oliveira HM, Polvikoski T, et al. TDP-43 proteinopathies: a new wave of neurodegenerative diseases. J Neurol Neurosurg Psychiatry. 2020 Nov;92(1):86–95.
20. Chevalier E, Audrain M, Ratnam M, Ollier R, Fuchs A, Piorkowska K, et al. Targeting the TDP-43 low complexity domain blocks spreading of pathology in a mouse model of ALS/FTD. Acta Neuropathol Commun. 2024 Oct;12(1):156.
21. Jo M, Lee S, Jeon YM, Kim S, Kwon Y, Kim HJ. The role of TDP-43 propagation in neurodegenerative diseases: integrating insights from clinical and experimental studies. Exp Mol Med. 2020 Oct;52(10):1652–62.
22. Wang X, Hu Y, Xu R. The pathogenic mechanism of TAR DNA-binding protein 43 (TDP-43) in amyotrophic lateral sclerosis. Neural Regen Res. 2024 Apr;19(4):800–6.
23. Floris G, Borghero G, Cannas A, Di Stefano F, Murru MR, Corongiu D, et al. Clinical phenotypes and radiological findings in frontotemporal dementia related to TARDBP mutations. J Neurol. 2015 Feb;262(2):375–84.
24. Caroppo P, Prioni S, Maderna E, Grisoli M, Rossi G. New MAPT variant in a FTD patient with Alzheimer's disease phenotype at onset. Neurol Sci. Italy; 202; 42:2111–4.
25. Domoto-Reilly K, Davis MY, Keene CD, Bird TD. Unusually long duration and delayed penetrance in a family with FTD and mutation in MAPT (V337M). Am J Med Genet B Neuropsychiatr Genet. 2017 Jan;174(1):70–4.
26. Zhang CC, Xing A, Tan MS, Tan L, Yu JT. The role of MAPT in neurodegenerative diseases: genetics, mechanisms and therapy. Mol Neurobiol. 2016 Sep;53(7):4893–904.
27. Koppers M, Blokhuis AM, Westeneng HJ, Terpstra ML, Zundel CAC, Vieira de Sá R, et al. C9orf72 ablation in mice does not cause motor neuron degeneration or motor deficits. Ann Neurol. 2015 Sep;78(3):426–38.
28. Chew J, Gendron TF, Prudencio M, Sasaguri H, Zhang YJ, Castanedes-Casey M, et al. Neurodegeneration. C9ORF72 repeat expansions in mice cause TDP-43 pathology, neuronal loss, and behavioral deficits. Science. 2015 Jun;348(6239):1151–4.
29. Liu Y, Pattamatta A, Zu T, Reid T, Bardhi O, Borchelt DR, et al. C9orf72 BAC mouse model with motor deficits and neurodegenerative features of ALS/FTD. Neuron. 2016 May;90(3):521–34.

5 Rodent Models of Lewy Body Disease

Fariba Akhondzadeh

5.1 INTRODUCTION

Lewy body disease (LBD) is a progressive neurodegenerative disorder characterized by the abnormal accumulation of misfolded α-synuclein within neurons, forming eosinophilic cytoplasmic inclusions known as Lewy bodies and Lewy neurites. In addition to α-synuclein, these inclusions contain other proteins, such as ubiquitin, neurofilament proteins, and synphilin-1, as well as lipids. First described in 1912 by the neurologist Friedrich Heinrich Lewy, these pathological aggregates are now recognized as a hallmark of LBD and related synucleinopathies.

The presence of α-synuclein aggregates in key brain regions including the substantia nigra, limbic system, and neocortex contributes to widespread neurodegeneration and dysfunction of multiple neurotransmitter systems, particularly the dopaminergic and cholinergic pathways. Despite extensive research, the precise mechanisms underlying LBD pathogenesis remain unclear. The disease is thought to arise from a complex interplay of genetic and environmental factors, oxidative stress, mitochondrial dysfunction, and neuroinflammation.

Rodent models are essential for unraveling the mechanisms of LBD and assessing potential therapeutic strategies. These models fall into two primary categories: genetic and pharmacological, each offering unique advantages in studying neurodegenerative disease processes. Genetic models provide valuable insights into the contributions of specific genes to disease onset and progression, while pharmacological models enable the rapid induction of pathology, facilitating the evaluation of potential treatments. Given the complexity of LBD, an integrative approach that combines both genetic and pharmacological models is often necessary to accurately replicate disease features and advance the development of effective therapies.

This chapter provides an overview of the pathological hallmarks of LBD and introduces existing rodent models used to study its underlying mechanisms. A deeper understanding of these models is essential for developing effective therapies and advancing research into α-synucleinopathies.

5.2 LEWY BODY DISEASE PATHOLOGY

Lewy bodies and Lewy neurites disrupt cellular homeostasis, leading to widespread neurodegeneration. The formation of Lewy bodies begins with the misfolding and aggregation of α-synuclein, a 140-amino-acid protein encoded by the SNCA gene. Normally, α-synuclein plays a role in synaptic vesicle regulation and neurotransmitter release, but under pathological conditions such as genetic mutations (A53T, A30P), oxidative stress, or post-translational modifications like phosphorylation and ubiquitination, it misfolds and forms insoluble fibrils, which interfere with neuronal function by impairing synaptic transmission, mitochondrial activity, and protein degradation pathways, ultimately leading to cell death.

The propagation of Lewy body pathology follows a prion-like mechanism, where misfolded α-synuclein spreads from cell to cell through exosomes or direct transmission, leading to the progression of disease across different brain regions. Several factors contribute to Lewy body formation, including genetic, cellular, and environmental influences.

DOI: 10.1201/9781032676524-6

TABLE 5.1
Pharmacological and Genetic Models of Lewy Body Disease in Rodents

Lewy Body Inducers		Mechanism of Action
Pharmacological models	Alpha-synuclein preformed fibrils (PFFs)	A specific fibrillar conformation of α-synuclein
	Rotenone	Mitochondrial complex I inhibitor
	Paraquat & Maneb	Environmental toxin leads to oxidative stress
	MPTP	Taken up by dopaminergic neurons, mimicking dopamine
	Lipopolysaccharide (LPS)	Neuroinflammation
	MG-132	Proteasome inhibitors
	Lactacystin	Proteasome inhibitors
	Tunicamycin	ER stress inducer
	Thapsigargin	ER stress inducer
Genetic models	Alpha-synuclein transgenic mice (SNCA-overexpression)	N/A

Genetic mutations in *SNCA* directly enhance α-synuclein's aggregation propensity, while mutations in genes such as *LRRK2, PARKIN,* and *PINK1* indirectly contribute by disrupting protein degradation and mitochondrial function. Cellular factors, including impairments in the ubiquitin-proteasome system (UPS) and autophagy-lysosome pathway, prevent the clearance of misfolded α-synuclein, allowing it to accumulate and form toxic aggregates. Mitochondrial dysfunction and oxidative stress further exacerbate this process by increasing reactive oxygen species (ROS) production, which promotes protein misfolding and lipid peroxidation. Additionally, interactions between α-synuclein and lipid membranes, particularly under conditions of oxidative damage, enhance its aggregation, contributing to Lewy body pathology.

Environmental factors also play a crucial role in disease development. Exposure to environmental toxins such as pesticides and heavy metals has been associated with an increased risk of α-synuclein aggregation. Traumatic brain injury (TBI) can trigger early aggregation events, increasing susceptibility to Lewy body pathology. Furthermore, chronic neuroinflammation, marked by activated microglia and elevated levels of pro-inflammatory cytokines, accelerates α-synuclein aggregation and exacerbates neuronal loss. Together, these factors drive the progressive neurodegeneration seen in Lewy body-associated diseases. Understanding these mechanisms provides critical insights into disease pathology and highlights potential therapeutic targets for intervention (1). The summarized pharmacological and genetic models of Lewy body disease in rodents are presented in Table 5.1.

5.3 RODENT MODELS OF LEWY BODY DISEASE

5.3.1 Non-Genetic Models

Pharmacological models involve administering drugs or neurotoxins to induce or modify Lewy body pathology and study the effects on disease mechanisms and symptoms.

5.3.1.1 Alpha-Synuclein Preformed Fibrils (PFFs)

Alpha-synuclein preformed fibrils (PFFs) represent a specific fibrillar conformation of α-synuclein that plays a crucial role in facilitating the misfolding and aggregation of endogenous α-synuclein monomers. In experimental models, PFFs exhibit prion-like properties, as they induce native α-synuclein to adopt an abnormal conformation, leading to the formation of insoluble aggregates. These aggregates closely resemble Lewy body pathology observed in human neurodegenerative diseases.

A key characteristic of PFFs is their ability to propagate trans-synaptically within neuronal circuits, enabling the progressive spread of pathology across different brain regions. In studies utilizing PFFs to model disease, various injection sites have been employed, including the striatum, substantia nigra, hippocampus, and cortex (2,3). Dopaminergic neurodegeneration (4), synaptic dysfunction (5), and motor impairments, along with cognitive impairments(6) have been observed following PFFs injections.

5.3.1.2 Rotenone

Rotenone, a widely used pesticide, is a mitochondrial complex I inhibitor known to induce oxidative stress and α-synuclein aggregation. Systemic administration of rotenone in rodents effectively models key pathological features of Lewy body disorders, including progressive neurodegeneration (7–9).

5.3.1.3 Paraquat and Maneb

Co-administration of these environmental toxins induces oxidative stress and promotes α-synuclein aggregation (10). It is important to note that the combined exposure to paraquat and maneb has been found to have a more pronounced effect on neurodegeneration than exposure to either chemical alone (11).

5.3.1.4 1-Methyl-4-Phenyl-1,2,3,6-Tetrahydropyridine (MPTP)

1-Methyl-4-phenyl-1,2,3,6-tetrahydropyridine (MPTP) is a neurotoxin commonly used to induce PD-like pathology, including the formation of Lewy body-like inclusions, in animal models. Upon metabolism in the brain, MPTP is converted into 1-methyl-4-phenylpyridinium (MPP+), a compound that selectively targets and damages dopaminergic neurons in the substantia nigra. This results in motor deficits and α-synuclein aggregation, characteristic of LBDs. The mechanism of MPTP toxicity is closely linked to its ability to mimic dopamine, allowing it to be taken up by neurons, where it disrupts mitochondrial function, induces oxidative stress, promotes α-synuclein misfolding, leads to Lewy body formation, and ultimately causes neuronal cell death.

5.3.1.5 Lipopolysaccharide (LPS)

Lipopolysaccharide (LPS) is a bacterial endotoxin frequently used to induce neuroinflammation and secondary α-synuclein aggregation in animal models. It is typically administered either intracerebrally or systemically.

5.3.1.6 Carbobenzoxy-Leu-Leu-Leucinal (MG-132)

MG-132 is a potent proteasome inhibitor that disrupts the UPS, leading to the accumulation of misfolded proteins, including α-synuclein. It is commonly administered intracerebrally.

5.3.1.7 Lactacystin

Lactacystin is a microbial metabolite that irreversibly inhibits the proteasome, leading to α-synuclein aggregation. Like MG-132, lactacystin is mostly administered intracerebrally (12).

5.3.1.8 Tunicamycin

Tunicamycin is an antibiotic that inhibits N-linked glycosylation in the endoplasmic reticulum (ER), leading to the accumulation of misfolded proteins, such as α-synuclein, and subsequent ER stress. This compound is administered in two ways to induce models: systemically and intracerebrally (13,14).

5.3.1.9 Thapsigargin

Thapsigargin is a plant-derived sesquiterpene lactone that inhibits the sarco/endoplasmic reticulum Ca^{2+}-ATPase (SERCA), disrupting calcium homeostasis and inducing ER stress. This disruption leads to the accumulation of misfolded proteins, such as α-synuclein, further exacerbating ER stress. Thapsigargin is primarily administered intracerebrally, with systemic administration being less common (15).

5.3.2 Genetic Models

Genetic manipulation for alpha-synuclein overexpression is a widely used method to study Lewy body pathology in animal models. This approach involves creating transgenic animals, typically mice or rats, that carry human alpha-synuclein genes with or without mutations or alpha-synuclein protein clearance gene knockout animals.

5.3.2.1 Alpha-Synuclein Transgenic Mice

5.3.2.1.1 SNCA-Overexpressing Transgenic Mouse Model

Transgenic mouse models that overexpress various forms of SNCA tend to develop α-synuclein aggregates over time, closely resembling the progressive nature of synucleinopathy. While these models typically do not exhibit neurodegeneration as severe as that seen in toxin-based models, they provide valuable insight into the genetic aspects of the disease. Specifically, they mimic the pathogenic mechanisms observed in humans with SNCA gene mutations or triplications, where overexpression of α-synuclein drives disease progression. One of the most widely used transgenic models of synucleinopathy is the M83 line, which overexpresses the A53T mutant form of human SNCA, driven by the prion promoter. The line 61 transgenic mouse model, which expresses full-length wild-type human SNCA under the Thy-1 promoter, is another well-established model of synucleinopathy. These mice show the following (9,16,17):

- Age-dependent locomotor changes
- Cholinergic deficits
- Cognitive impairment
- Fine- and sensorimotor impairments
- Inflammation
- Synaptic deficits

REFERENCES

1. Fares MB, Jagannath S, Lashuel HA. Reverse engineering Lewy bodies: how far have we come and how far can we go? Nat Rev Neurosci. 2021;22(2):111–31.
2. Jin JW, Fan X, del Cid-Pellitero E, Liu XX, Zhou L, Dai C, et al. Development of an α-synuclein knockdown peptide and evaluation of its efficacy in Parkinson's disease models. Commun Biol. 2021;4(1):232.
3. Patterson JR, Kochmanski J, Stoll AC, Kubik M, Kemp CJ, Duffy MF, et al. Transcriptomic profiling of early synucleinopathy in rats induced with preformed fibrils. NPJ Parkinsons Dis. 2024;10(1):7.
4. Abdelmotilib H, Maltbie T, Delic V, Liu Z, Hu X, Fraser KB, et al. α-Synuclein fibril-induced inclusion spread in rats and mice correlates with dopaminergic Neurodegeneration. Neurobiol Dis. 2017 Sep;105:84–98.
5. Volpicelli-Daley LA, Luk KC, Patel TP, Tanik SA, Riddle DM, Stieber A, et al. Exogenous α-synuclein fibrils induce Lewy body pathology leading to synaptic dysfunction and neuron death. Neuron. 2011 Oct;72(1):57–71.
6. Kim S, Kwon SH, Kam TI, Panicker N, Karuppagounder SS, Lee S, et al. Transneuronal propagation of pathologic α-synuclein from the gut to the brain models Parkinson's disease. Neuron. 2019 Aug;103(4):627–641.e7.

7. Pan-Montojo F, Anichtchik O, Dening Y, Knells L, Pursche S, Jung R, et al. Progression of Parkinson's disease pathology is reproduced by intragastric administration of rotenone in mice. Nature Precedings. 2010;1.
8. Landau R, Halperin R, Sullivan P, Zibly Z, Leibowitz A, Goldstein DS, et al. The rat rotenone model reproduces the abnormal pattern of central catecholamine metabolism found in Parkinson's disease. Dis Model Mech. 2022 Jan;15(1).
9. Paumier KL, Sukoff Rizzo SJ, Berger Z, Chen Y, Gonzales C, Kaftan E, et al. Behavioral characterization of A53T mice reveals early and late stage deficits related to Parkinson's disease. PLoS One. 2013;8(8):e70274.
10. Che Y, Hou L, Sun F, Zhang C, Liu X, Piao F, et al. Taurine protects dopaminergic neurons in a mouse Parkinson's disease model through inhibition of microglial M1 polarization. Cell Death Dis. 2018;9(4):435.
11. Anselmi L, Bove C, Coleman FH, Le K, Subramanian MP, Venkiteswaran K, et al. Ingestion of subthreshold doses of environmental toxins induces ascending Parkinsonism in the rat. NPJ Parkinsons Dis. 2018;4(1):30.
12. Bentea E, Van der Perren A, Van Liefferinge J, El Arfani A, Albertini G, Demuyser T, et al. Nigral proteasome inhibition in mice leads to motor and non-motor deficits and increased expression of Ser129 phosphorylated α-synuclein. Front Behav Neurosci. 2015;9:68.
13. Wang H, Wang X, Ke ZJ, Comer AL, Xu M, Frank JA, et al. Tunicamycin-induced unfolded protein response in the developing mouse brain. Toxicol Appl Pharmacol. 2015;283(3):157–67.
14. Han X, Zhang X, Li H, Huang S, Zhang S, Wang F, et al. Tunicamycin enhances the antitumor activity of trastuzumab on breast cancer in vitro and in vivo. Oncotarget. 2015;6(36):38912.
15. Abdullahi A, Stanojcic M, Parousis A, Patsouris D, Jeschke MG. Modeling acute ER stress in vivo and in vitro. Shock. 2017;47(4):506–13.
16. Chesselet MF, Richter F, Zhu C, Magen I, Watson MB, Subramaniam SR. A progressive mouse model of Parkinson's disease: the Thy1-aSyn ("Line 61") mice. Neurotherapeutics. 2012 Apr;9(2):297–314.
17. Subramaniam SR, Magen I, Bove N, Zhu C, Lemesre V, Dutta G, et al. Chronic nicotine improves cognitive and social impairment in mice overexpressing wild type α-synuclein. Neurobiol Dis. 2018 Sep;117:170–80.

6 Rodent Models of Huntington's Disease

Bita Firouzan

6.1 INTRODUCTION

Huntington's disease (HD) is a devastating, inherited neurodegenerative disorder characterized by progressive motor impairments, cognitive decline, and psychiatric manifestations, ultimately leading to death (1). The disease follows an unrelenting course, with patients typically succumbing to complications such as aspiration pneumonia, dysphagia, suicide, or general systemic exhaustion within 15–20 years of clinical diagnosis. The global prevalence of HD is estimated to be approximately five to eight cases per 100,000 individuals (2).

First described by George Huntington in 1872, HD is named in his honor. It is an autosomal dominant nervous system disorder caused by an unsteady expansion of CAG repeat within the coding region (first exon) of the huntingtin (htt) gene, located on chromosome 4p16.3, which triggers widespread neurodegeneration. While HD predominantly manifests in adults between 30 and 50 years of age, cases of both juvenile-onset (before age 20) and late-onset (after age 60) have been reported, indicating a broader age spectrum of disease presentation (1).

While no animal model fully recapitulates the human disease condition, a significant portion of our current understanding of disease pathogenesis is derived from animal models. Therefore, they serve as valuable tools in scientific research, and preclinical studies profoundly rely on them. Animal models can recapitulate signs and symptoms observed in humans by employing specific neurotoxins or through genetic engineering. The selection of an animal model largely depends on the specific experimental question.

A variety of vertebrate and invertebrate models of HD have been established to date, including *Caenorhabditis elegans* (*C. elegans*), *Drosophila*, zebrafish, rodents, and larger animal models such as pigs, sheep, and rhesus monkeys. Among these, rodent (mouse, rat) models are the most widely used and more accurately represent the progression and pathogenesis of HD (3–6). The majority of animal models of HD can be classified into two main categories: genetic and non-genetic (toxin-based) models. Pioneering HD studies mostly used neurotoxins, such as Quinolinic Acid (QA) and 3-Nitropropionic acid (3-NP) to generate non-genetic models. However, with the discovery of the htt protein and the associated mutation in 1993, various rodent genetic (transgene and knock-in (KI)) models have been developed (6–9). A valid rodent model of HD should recapitulate, to the extent possible, the neuropathological and clinical features of the disease. Therefore, in this chapter, we will first explore the key pathological hallmarks of HD, including its molecular and cellular mechanisms. Following this, we aim to provide an overview of the most extensively studied and available rodent models of HD that can aid in therapeutic drug discovery and development in the future. We will also discuss the advantages and pitfalls of these models in detail. It is important to note that the models described in this chapter do not represent a comprehensive list.

6.2 PATHOGENESIS AND CLINICAL FEATURES OF HUNTINGTON'S DISEASE

The htt gene, also known as the IT15 gene, encodes a ubiquitously expressed large protein known as huntingtin, and the CAG trinucleotide corresponds to the amino acid (a.a) glutamine. The 350-kDa htt protein is highly expressed throughout the brain. In 1993, researchers discovered that

 DOI: 10.1201/9781032676524-7

mutations in the htt gene result in the formation of a polyglutamine (polyQ) domain in the N-terminal region of the htt protein (10,11). Abnormal htt protein, along with other proteins, can aggregate and form intranuclear and cytoplasmic inclusions, leading to neuronal demise, particularly within the basal ganglia circuits (12–14). However, there is controversy surrounding the role of intracellular inclusions in neuronal degeneration, as evidence supports both neuroprotective and harmful effects (15–17). It is important to point out that a significant inverse correlation exists between the length of the CAG triplet repeat and the age of disease onset (18–20). In general, an intermediate range of 27–35 repeats is considered normal. However, the expansion of trinucleotide repeats of 36 or more (≥36) is deemed pathogenic. Elongated repeats in htt typically lead to an earlier onset of the disease (21,22). Despite this factor, researchers have identified additional causative modifiers that play a determining role in the clinical onset of HD (23,24).

The progressive degeneration of neuronal cells occurs in the brain of HD, initially in the striatum, followed by other regions such as the cerebral cortex, hypothalamus, cerebellum, thalamus, substantia nigra, hippocampus, and brain stem nuclei (25–27). Wild-type (WT) htt predominantly localizes in the cytoplasm of cells, and its biological functions are not yet fully understood. However, growing evidence supports significant roles for the htt protein in regulating gene transcription, intracellular transport, cell survival, trafficking of synaptic vesicles, and autophagy (28–30). Furthermore, mutant htt (m-htt) interferes with mitochondrial function (31,32) and immune activity (33), as reported in HD patients.

The neuropathological hallmark of HD is the prevalent degeneration of GABAergic medium spiny neurons in the striatum, a brain region crucial for movement control. This neuronal population is particularly vulnerable to m-htt, and the neurodegeneration leads to a characteristic pattern of atrophy in the caudate nucleus and putamen (34,35), which is believed to be linked to one of the well-known motor symptoms of HD: *Chorea*, irregular, uncontrolled movements that typically affect the trunk, face, and extremities (36). From the viewpoint of a movement disorder, additional motor symptoms of HD encompass dystonia, akinesia, bradykinesia, rigidity, dysphagia, dysarthria, gait irregularities, incoordination, and postural instability that insidiously worsen with time (37). Although the clinical diagnosis of HD onset is primarily based on motor manifestations, cognitive and psychiatric symptoms are prominent among HD patients and most often precede the emergence of motor symptoms (38–40). Hence, recognizing these non-motor signs may enable an early diagnosis of HD, providing a timely opportunity for neuroprotective interventions, which are deemed more effective in the earlier stages of a progressive nervous system disorder.

Cognitive alterations in HD may include personality changes, executive dysfunction, difficulties in organizing and planning, as well as deficits in attention and working memory (39,41,42). The third main category of clinical symptoms of HD includes psychiatric disturbances, such as depression, anxiety, suicidal thoughts, apathy, irritability, and aggression, which are prevalent among HD patients (43,44). In addition, weight loss, sleep disturbances, and circadian rhythm disruptions are noted in individuals with HD (45,46). The non-motor symptoms of HD can significantly compromise patients' quality of life, just as the motor symptoms do.

Despite extensive research over the past few decades, the disease remains incurable, and the available treatments merely alleviate symptoms, delay death, and do not change the natural course of the disease. A crucial step in identifying effective drugs is clarifying the pathophysiological mechanisms underlying the disease. In this context, animal models play an essential role in exploring the pathogenesis of HD and offering insights into the outcomes of putative therapeutic approaches.

6.3 RODENT MODELS OF HUNTINGTON'S DISEASE

6.3.1 Non-Genetic Models

Prior to the discovery of the genetic mutation responsible for HD, rodent chemical models were developed using neurotoxins based on two phenomena observed in HD brains: excitotoxicity and mitochondrial dysfunction, both of which induce neuronal cell death (6). These initial non-genetic models played a crucial role in unraveling the underlying mechanisms of HD pathogenesis. Here, we focus on rodent neurotoxin-induced models of HD.

6.3.1.1 Excitotoxic Models

As mentioned earlier, while GABAergic MSNs in the striatum primarily undergo massive degeneration in HD (47), striatal interneurons are relatively preserved from neurodegeneration. One possible reason for the selective loss of the MSN population may be attributed to N-methyl-d-aspartate (NMDA) receptor-mediated excitotoxicity (48). It is well-known that striatal MSNs receive glutamatergic afferents from the cortico-striatal pathway, and they express high levels of NMDA receptors, rendering them more vulnerable to excitotoxicity (49). During the excitotoxicity process, over-activation of NMDA receptors leads to alterations in intracellular Ca^{2+} concentrations, mitochondrial dysfunction, and oxidative damage, ultimately resulting in neuronal cell death (50). Moreover, several lines of studies have demonstrated that excitotoxicity induces selective striatal cell loss due to increased glutamate neurotransmission (51). Consequently, in the late 1970s, the first rodent model employing the excitotoxic agent, kainic acid (KA) to destroy MSNs in the striatum was developed (52,53). Then, researchers used another excitotoxin, quinolinic acid (QA), which acts as a selective NMDA receptor agonist and exhibits pathological features of HD better than the KA models. The QA toxic mechanism is responsible for selective damage of striatal MSNs and motor impairments in affected animals, closely mirroring the changes observed in HD (54). QA is an endogenous downstream metabolite generated from tryptophan metabolism through the kynurenine pathway. Under normal physiological conditions, QA is not harmful. However, only a slight increase in its levels causes cell damage (55). Postmortem studies of HD patients have shown elevated activity of the enzyme 3-hydroxyanthranilic acid oxygenase (3-HAO), as well as increased levels of QA in the striatum (56,57). This enzyme converts 3-hydroxyanthranilic acid, an intermediate in the kynurenine pathway, to QA. In animal models, QA must be directly injected into the striatum, as it cannot cross the blood-brain barrier (BBB). Following injection, it induces the death of striatal MSNs in mice, rats, and non-human primates, resembling some of the main characteristics of neurodegeneration observed in HD brains (9,58,59). Additionally, QA injection into the mouse striatum stimulates the upregulation of the htt protein (6). QA-induced rodent models exhibit motor and cognitive impairments. Motor abnormalities, such as hyperactivity, resemble the early stages of HD. In addition, unilateral injection of QA in rodents results in apomorphine-induced rotations (60). Furthermore, rats that underwent unilateral QA treatment tend to use their ipsilateral forelimb for general activities due to the impairment of their contralateral forelimbs (61). Cognitive symptoms in rodent QA models include deficits in procedural memory, visuospatial skills, and poor memory recall, which were assessed using the T-maze, Morris water maze, and radial arm water maze (62–64).

6.3.1.2 Metabolic Models

Another mechanism contributing to neuronal cell death in individuals with HD is mitochondrial dysfunction and related deficits in energy metabolism (50,65). In HD brains, a downregulation of several tricarboxylic acid cycle (TCA) enzymes and a reduction in the activity of mitochondrial enzymatic complexes II, III, and IV have been reported (66,67). Disruption of mitochondrial enzyme activity involved in the electron transport chain is linked to the abnormal increase in the production of reactive oxygen species (ROS), including hydrogen peroxide, superoxide radicals, and hydroxyl radicals. ROS can harm the cell membrane and genetic material (66). Additionally, postmortem studies indicated that adenosine triphosphate (ATP) production by mitochondria is reduced in HD brain samples compared to healthy controls (31). Furthermore, the ultrastructure, biogenesis, dynamics, and enzyme complexes activity of mitochondria are altered in HD brains (66–68). Remarkably, the mhtt protein is involved in all these pathways (48). In this context, several mitochondrial inhibitors have been applied to recapitulate striatal degeneration and similar movement anomalies observed in individuals with HD, including 3-NP, malonate, Aminooxy acetic acid (AOAA), and succinate (69–71). Among these, 3-NP is believed to establish a reliable HD animal model.

3-NP is a neurotoxin that irreversibly induces mitochondrial impairment by suppressing succinate dehydrogenase, a mitochondrial enzyme crucial for the respiratory chain (72). The 3-NP toxin leads to selective striatal damage, particularly in the lateral region, by causing degeneration of GABAergic MSNs (9). This mitochondrial toxin can cross the BBB. Therefore, it can be administered systemically to rodents and non-human primates. It is noteworthy that specific rat strains show significantly different responses to 3-NP (73).

The 3-NP models indicate motor dysfunction and cognitive decline. Depending on the timing of injection in rats, 3-NP enables it to resemble both the hypokinetic and hyperkinetic features of HD (74). In rats, 3-NP can influence their working and reference memories and impair habituation (75).

6.3.1.3 Advantages and Pitfalls of Non-Genetic Models

There is currently no comprehensive model for studying HD that replicates all aspects of the condition in humans. However, toxin-based models can serve as a platform for evaluating the efficacy of various therapeutic interventions aimed at neuroprotection or disease modification. These models offer several advantages, such as easy accessibility and the straightforward administration of drugs to both rodents and non-human primates, which can induce significant and predictable neuronal cell loss in specific brain regions, such as the striatum, which is significantly affected in HD. Neurotoxin administration can quickly induce neurological deficits, enabling researchers to study the effects of neurodegeneration and potential treatments in a relatively short timeframe. By specifically causing neurodegeneration, these models can help elucidate the biochemical and molecular pathways involved in neuronal death and dysfunction closely related to HD. Unlike genetic models, toxin-based models do not rely on specific genetic modifications, which can sometimes introduce variability in disease phenotypes. This can make the effects of neurotoxins more straightforward to interpret.

While toxin-based models of HD offer significant advantages for studying neurodegeneration and testing treatments, they also present limitations that researchers must consider when designing studies and interpreting findings. Neurotoxin models do not exhibit the genetic basis of HD, which is characterized by CAG repeat expansions in the htt gene. This limits their ability to model the hereditary aspects of the disease and its progression. The effects of neurotoxin administration are often acute and do not replicate the progressive nature of HD, where over time there is a gradual degeneration of neurons and a range of behavioral and cognitive symptoms. Hence, these models might not fully capture the complete range of symptoms associated with HD, especially those related to cognitive and psychiatric disturbances, which are significant components of the human condition.

6.3.2 Genetic Models

HD is a monogenic neurodegenerative disorder, and so far, there has been no effective treatment for those affected by HD. The discovery of the genetic association between HD and the htt gene has led to the development of rodent genetic models to illuminate the underlying pathological mechanisms by which m-htt induces neuronal cell death. To date, various rodent genetic models have been created, each exhibiting a range of HD-like features. Nonetheless, none of these models demonstrates the significant neuronal loss characteristic of human HD brains. In this context, two main categories are transgenic and KI models. The following section will review the constructs and phenotypes of the commonly used rodent genetic models.

6.3.2.1 Transgenic Models

Transgenic rodent models are genetically engineered animals, typically mice (*Mus musculus*) or rats (*Rattus norvegicus*), that contain an exogenous DNA (truncated or full-length) introduced into their genome. This modification enables researchers to study the functions and effects of specific genes, explore disease mechanisms, and assess the impact of genetic alterations in a controlled environment. These models are important tools for elucidating the pathogenesis of HD and paving the way for the development of potential treatment strategies.

6.3.2.1.1 Fragmented Transgenic Mouse Models

6.3.2.1.1.1 R6/1 and R6/2 To model certain features of HD, the first transgenic mouse lines, known as R6, were created by Bates and colleagues in 1996. The R6 transgenic lines (R6/1 and R/6/2) express a truncated N-terminal htt fragment with an expanded CAG repeat, controlled by the human htt promoter, which is randomly inserted into the mouse genome. Table 6.1 summarizes some key features of each case. Structurally, R6/1 and R6/2 were generated using a similar method. However, they vary in CAG repeat length, with R6/1 containing ~115 CAG repeats and R6/2 carrying ~145 CAG repeats (76,77). Despite expressing the human m-htt gene, transgenic models also express two WT copies of their own htt gene. These models were developed to study the impact of the m-htt protein associated with HD and to evaluate potential treatment approaches. Both models resemble progressive phenotypes of HD. In general, R6/1 mice show relatively milder behavioral phenotypes compared to R6/2 mice, which may be related to their lower CAG repeat length. Furthermore, the transgenic expression level in the R6/1 model is approximately 31%, while in the R6/2 model, it is nearly 75% (76).

R6/1 mice display striatal shrinkage without any apparent cell death. However, they possess neuronal intranuclear inclusions (NIIs) (77). In addition, this model shows body weight loss, reduced motor activity, hindlimb clasping, gait abnormalities, cognitive deficits, tremors, occasional epileptic seizures, anxiety- and depressive-like behaviors, shortened life span (76,78–82), as well as

TABLE 6.1
Transgenic Rodent Models of Huntington's Disease

Genetic Model	Host Animal	PolyQ Length (CAG)	Strain Background	Construct	Promoter	Striatal Atrophy	NIIs Spread	Lifespan
R6/1	Mouse	~115	C57BL/6J, BALB/cByJ, B6CBA/Ca mixed	Fragment (exon 1 of the human htt gene)	Human htt promoter	+	Throughout	Reduced (~9 M)
R6/2	Mouse	~145	C57BL/6J, DBA/2J, B6CBA/Ca mixed	Fragment (exon 1 of the human htt gene)	Human htt promoter	+	Throughout	Reduced (13–15 W)
N171-82Q	Mouse	82	C57BL/6× B6C3H/He J mixed	Fragment (first 171 a.a of human htt gene)	Mouse prion promoter	+	Throughout	Reduced (6–11 M)
YAC128	Mouse	128	FVB/N C57BL/6J	Full-length	Human htt promoter	+	Throughout	Relatively normal
BACHD	Mouse	97	FVB/N	Full-length	Human htt promoter	+	Striatum Cortex	Relatively normal
tgHD rat	Rat	51	Sprague-Dawley	Fragment	Rat htt promoter	+	Throughout	Sorter than WT
BACHD rat	Rat	97	Sprague-Dawley	Full-length	Human htt promoter	+	Throughout	Sorter than WT

W = weeks, M = months, a.a = amino acid, WT= wilde-type

aberrant synaptic plasticity (83). Symptoms typically appear at approximately 4.5 months of age and gradually deteriorate over time (79).

R6/2 is the most commonly studied model of HD and exhibits a more robust and aggressive phenotype profile than R6/1 mice. Compared to other genetic models, R6/2 has a rapid disease onset and progression, similar to juvenile HD in humans. As a result, R6/2 serves as an excellent model for exploring the pathogenesis of juvenile HD and for preclinical pharmacology (84). Neurological symptoms may appear at almost four weeks of age, and animals only survive for 13–15 weeks. R6/2 mice experience progressive weight loss, brain volume reduction, striatal cell loss, and atrophy (76,77,85–87). Motor incoordination, abnormal gait, and procedural learning deficits are among the pronounced symptoms (76,88). Like R6/1, this model shows anxiety- and depressive-like phenotypes (89,90). They also exhibit marked NIIs during the prodromal phase, which progressively distribute in several brain regions, such as the striatum, cortex, and hippocampus (77,91,92).

As in human HD, transgenic R6/2 mice exhibit overt motor impairments and cognitive decline. In this regard, resting tremor, hypokinesis, dyskinesia, clasping behavior, epileptic seizures, poor rotarod, and beam walking performance and diminished grip strength are associated with motor disturbances (76,78,85,86). Furthermore, R6/2 mice display an intense loss of orexin-expressing neurons in the hypothalamus, akin to those observed in individuals with HD (93). Orexin is associated with energy metabolism, the sleep-wake cycle, and emotion (94). Dysregulation of gene transcription (95), changes in signaling pathways (96), cardiac dysfunction (97), and skeletal muscle abnormalities (98) are also detectable in this model. Cognitive decline, including spatial learning deficits assessed by the Morris water maze test, is reported at 3.5 weeks of age. R6/2 mice also demonstrate poor performance on the alternating T-maze task (78) and show anomalies in fear conditioning tests and exploratory activity (99).

There is evidence of alterations in the expression of various neurotransmitter receptors in R6/2 mice. For example, dopamine D_1 and D_2 receptors in striatal projection neurons are remarkably reduced, along with other neurotransmitter receptors, including glutamate metabotropic receptors ($mGluR_1$, $mGluR_2$, and $mGluR_3$) (100). Moreover, abnormal synaptic plasticity at the CA_1 region of the hippocampus was reported before the overt appearance of symptoms (101).

6.3.2.1.1.2 N171-82q Another well-characterized truncated HD mice model, known as N171-82Q, was developed by expressing a cDNA that encodes the first 171 a.as of the N-terminal fragment of the human htt protein, which carries 82 CAG repeats, under the regulation of a mouse prion promoter (102). The m-htt expression occurs throughout the mouse brain and is confined to neurons (not found in glia). Compared to R6/2 mice, this model has fewer polyQ repeats, resulting in a later onset of HD symptoms, which may render it a valuable candidate for modeling adult-onset HD. Furthermore, N171-82Q mice show less severe phenotypes than R6/2 mice (Table 6.1).

N171-82Q mice exhibit normal phenotypes until they reach about two months of age. After that, they display both behavioral and anatomical symptoms. Behavioral symptoms include tremors, clasping, hypoactivity, loss of coordination, poor rotarod performance, and gait abnormalities (6,90,103). By 14 weeks, deficits in working and reference memory (104) as well as impairments in motor learning (90) are detectable in this model. In HD brains, cognitive decline is thought to stem from impairments in fronto-striatal circuitry. However, previous studies suggest that the cognitive decline observed in transgenic mouse models may be linked to the presence of htt inclusions in the hippocampus rather than to fronto-striatal circuitry (6).

From a neuropathological perspective, striatal cell loss and reduced volume have been reported at the age of four months (105). In N171-82Q mice, progressive cortical atrophy along with ventricular enlargement has been demonstrated compared to their WT littermates (106). Like the R6 line, N171-82Q also reveals NIIs in several brain regions, including the striatum, cortex, and hippocampus (107,108). In addition, progressive weight loss, degenerative alterations in the hypothalamus (95,109), gene expression changes (95,110), and depressive-like phenotypes (90) have been demonstrated in the truncated model N171-82Q. Despite its short lifespan, this model normally survives longer than R6/2 mice.

6.3.2.1.2 Full-Length Transgenic Mouse Models

6.3.2.1.2.1 YAC128 and BACHD Two of the most popular and relevant mouse models for clinical trials of HD are Yac128 and BACHD. These transgenic models were created using yeast artificial chromosome (YAC) and bacterial artificial chromosome (BAC), which express the full-length human htt gene carrying 128 and 97 polyQ stretches, respectively, under the regulation of the human htt promoter (111,112) (Table 6.1). It is important to note that there are several other YAC mouse models, such as YAC18, YAC46, and YAC72, which have 18, 46, and 72 polyQ repeat expansions, respectively (113). However, we focus exclusively on YAC128 due to its ability to more robustly and earlier mimic the HD symptoms compared to the other YAC models. Moreover, among BAC models (BACHD and BAC-225Q), we will describe BACHD mice in detail.

Yac128 and BACHD models demonstrate germline or somatic stability, likely due to the existence of mixed CAG/CAA triplet repeats, both of which translate to glutamine (48). Notably, these mice also express two normal copies of their own htt gene. These full-length models recapitulate milder phenotypes and a slower disease progression compared to the fragment models. In contrast, they represent more pronounced phenotypes than KI mice. YAC128 and BACHD demonstrate striatal and cortical atrophy by 12 months of age (114,115). Neuropathological signs in the striatum appear prior to the cortex (114,116,117). Additionally, electrophysiological data indicate alterations in their glutamatergic synapses (118) as well as in the expression levels of glutamate receptors (119).

Despite many similarities, YAC128 and BACHD models differ in certain aspects. For example, in BACHD mice, htt aggregates are mainly present in the cortex rather than the striatum. Additionally, htt aggregate pathology predominantly develops in the cytoplasm. However, these aggregates appear to be less numerous than those found in Yac128 (111). Another difference is that YAC128 indicates neuronal cell loss in the striatum, whereas there are no reports of striatal cell loss in BACHD (111). Both models resemble motor and cognitive deficits from the age of three months (111,120,121). Anxiety and depressive-like behaviors have also been reported in these animals (89,90,121,122).

6.3.2.1.2.2 YAC128 YAC128 transgenic mice, generated at the Hayden laboratory, are the most extensively studied full-length model. In this model, progressive motor symptoms comprise a decline in rotarod performance, gait abnormalities, and early hyperactivity followed by late-stage hypoactivity, assessed using the open-field apparatus at the age of 12 months (89,114). Cognitive deficits in YAC128 mice that preceded the onset of motor symptoms have also been demonstrated (120). In this regard, Yac128 mice exhibited poor performance in a T-maze test and showed learning and memory impairments (120,123,124) compared to their WT littermates. Regarding neuropathological features, YAC128 mice display degeneration of striatal MSNs, preferentially in the lateral part (112,118). Loss of orexin (125), striatal NIIs (126,127), and increased body weight associated with htt over-expression (128) have also been reported. Unlike previous transgenic models, Yac128 has a relatively normal lifespan. Consequently, they are considered a valuable tool for long-lasting therapeutic strategies.

6.3.2.1.2.3 BACHD BACHD full-length mouse model was developed by Gray and colleagues in 2008 (111). Like the Yac128 phenotype profile, BACHD mice exhibit progressive motor deficits, striatal synaptic dysfunction, various learning impairments, and increased weight gain (84,111,129–131). Motor deficits include difficulties in rotarod performance as well as balance and gait abnormalities (89,112,114,132). Starting at six months onwards, BACHD mice demonstrate hypoactivity in the open-field test (133). In the case of learning impairments, this full-length model recapitulates deficits in reversal learning (120), motor learning (111), novel object recognition memory (131), and sensorimotor gating (89,120). Unlike Yac128, the increased body weight in BACHD mice is thought to be linked to higher food and water intake. Moreover, around two months of age, they display endocrine anomalies associated with an early obese phenotype (134). BACHD transgenic mice maintain a fairly normal life expectancy and display late-onset symptoms of the disease similar to YAC128. As a result, the full-length models discussed here have mostly been utilized to study HD characteristics at both pre-symptomatic and post-symptomatic phases.

6.3.2.1.3 Fragment Transgenic Rat Model of Huntington's Disease (tgHD)

The transgenic rat model (tgHD), characterized by von Horsten and colleagues, expresses a truncated htt cDNA fragment that contains 51 CAG triplet repeats under the control of the rat endogenous htt promoter, ultimately generating a human/rat combined protein (135) (Table 6.1). The expression of the m-htt gene is mostly detectable in the frontal and temporal cortices, BG, midbrain, hippocampus, with low expression levels in the spinal cord, and cerebellum. Generally, the tgHD model exhibits a progressive and late-onset disease phenotype (136). At birth, they seem normal, but gradually the symptoms become apparent as they age (137). The tgHD rats display severe weight loss at 24 months of age and maintain a shorter lifespan compared to their WT littermates (138). They also demonstrate motor abnormalities, including gait and balance anomalies, hyperactivity followed by hypoactivity as assessed by the open-field test, chorea-like movements, and a progressive decline in the rotarod apparatus (139–141). In 2008, Bode and colleagues stated that there are variations in behavioral phenotypes between male and female tgHD rats (142). Moreover, homozygous tgHD rats reveal more pronounced behavioral and neuropathological symptoms (139,141,143).

The non-motor symptoms are also evident in the tgHD model. Like HD patients, these symptoms typically manifest before motor symptoms and deteriorate over time. At both early and late stages of the disease, tgHD rats show cognitive decline that begins between six and nine months. This cognitive decline includes impaired acquisition in the water maze task, deficits in visuospatial memory, and decreased spatial working and reference memory (137,144–146). Furthermore, tgHD rats exhibit reduced anxiety-like behavior compared to WT rats in various behavioral tests (136,137,140).

A variety of neuropathological features have been documented in tgHD rats. Htt aggregates and nuclear inclusions are detected in the striatum, nucleus accumbens, and subsequently in the cortex and other brain regions, such as the thalamus and SNpc (136,137). As tgHD rats age, progressive striatal cell loss, profound cortical cell damage, striatal neuron shrinkage, and enlarged lateral ventricles, indicating striatal atrophy, have also been reported (139,147,148). However, some studies show controversy in the latter two cases (144,149,150).

6.3.2.1.4 Full-Length Transgenic Rat Model of Huntington's Disease (BACHD)

Like transgenic BACHD full-length mice, BACHD rats were developed using bacterial artificial chromosomes (BAC). They express the complete human htt gene with 97 CAG/CAA repeats, under the regulation of the human htt promoter (151) (Table 6.1). As this model has been developed recently, only limited information is currently available. However, BACHD rats exhibit motor symptoms, including impaired rotarod performance, imbalanced motor coordination, and disturbed walking gait (130,151). An assessment of locomotor activity through the open-field test revealed hyperactivity, which was then replaced by hypoactivity at four months of age, similar to the tgHD model (130). Regarding non-motor symptoms, these animals demonstrate a reduction in both anxiety-like and exploratory behaviors (151). From a neuropathological viewpoint, the BACHD rat model displays m-htt aggregates in the striatum, cortex, amygdala, nucleus accumbens, and hippocampus (151,152). Moreover, alterations in striatal gene expression have been noted in BACHD rats (153). Interestingly, male BACHD rats represent the obesity phenotype without indicating an increase in body weight. They even consume less food than WT rats (154). A reduced lifespan has not yet been reported in this model.

The advantages of using rats as rodent models lie in their generally longer lifespan compared to mice, which can be beneficial for studying the progression of chronic diseases like HD. This extended lifespan allows for the observation of long-term effects, therapeutic interventions, and late-stage symptoms. Moreover, rats possess a larger brain mass, making surgical procedures and manipulations easier and more precise. This is particularly advantageous for techniques like stereotaxic injections. In summary, while both rats and mice serve as valuable models for studying HD, rats may provide unique benefits in behavioral complexity, manipulation, and certain physiological traits that can enhance the understanding of the disease and its treatment.

6.3.2.2 Advantages and Pitfalls of Transgenic Models

Transgenic rodent models of HD have been instrumental in studying the disease's mechanisms and potential treatments. However, like any model system, they come with both advantages and limitations. These models often express mutant forms of the htt protein, mhtt, that are directly linked to human HD, allowing researchers to study the specific effects of the mutation in a controlled environment similar to that in humans. In addition, the models can replicate the progressive neurodegeneration observed in humans, which helps in understanding the temporal aspects of the disease and the effectiveness of potential therapeutic interventions over time. Many transgenic models display behavioral changes that mimic symptoms of HD, such as motor coordination deficits and cognitive impairments. This provides a platform for testing treatments aimed at ameliorating these symptoms. Transgenic mice exhibit considerable similarities to humans in terms of behavior, anatomy, and gene expression, yet they also display notable differences. HD is a complex condition influenced by various environmental, genetic, and epigenetic factors. While transgenic models can provide insights, they may oversimplify these interactions, resulting in findings that don't entirely apply to human patients. Additionally, transgenic mice typically exhibit an overexpression of the mhtt protein, which may not accurately represent the protein levels found in humans affected by the disease, potentially distorting outcomes. These mice possess two copies of the normal mouse htt gene and one copy of the human mhtt gene, whereas individuals with HD typically have one normal copy and one mutant copy of the htt gene. Different transgenic models may exhibit varying degrees of disease phenotypes, making it challenging to generalize results across studies unless the same model is used. Depending on the model, some transgenic rodents may not fully recapitulate all aspects of the human disease, such as certain non-motor symptoms, leading to gaps in understanding the disorder's full clinical presentation.

6.3.2.3 Knock-in Models

From a genetic viewpoint, KI mouse models are regarded as the most accurate models that faithfully replicate the human HD (84,155). To create this model, a portion of the endogenous mouse htt gene is substituted with a human m-htt that carries an expanded CAG repeat. Hence, KI mice possess only two copies of the htt gene (a mutant allele and a WT allele), and the expression of the m-htt gene is under the regulation of the mouse htt promoter (156). Generally, KI mice display a normal lifespan and a progressive, late-onset disease phenotype, making them suitable models for studying early neuropathological alterations associated with HD. However, they show milder pathology and behavioral abnormalities compared to previously described models (157–159). These genetic models recapitulate motor, cognitive, and psychiatric deficits related to HD, exhibiting varying degrees of severity based on the model (158,160,161). To date, several KI models featuring various expanded CAG repeats have been developed. The most extensively studied KI models include HdhQ111, CAG140, HdhQ150, and zQ175. Although none of them show overt striatal cell loss, striatal neuropathology is apparent (157).

6.3.2.3.1 HdhQ111

HdhQ111 mice, expressing 111 CAG repeats, demonstrate moderate pathology in the brain. Although they show no significant differences in behavioral pathology compared to their WT littermates, subtle motor dysfunction can be measured at 24 months of age. Heterozygous HdhQ111 mice show spatial memory deficits at eight months of age (160). Neuropsychiatric symptoms, including

TABLE 6.2
Knock-In Rodent Models of Huntington's Disease

Genetic Model	Host Animal	CAG Length	Strain Background	Construct	Promoter	Striatal Atrophy	NIIs Spread	Lifespan
HdhQ111	Mouse	111	C57BL/6JCBA	Full-length	Mouse htt promoter	–	Throughout	Normal
CAG140	Mouse	140	C57BL/6JCBA	Full-length	Mouse htt promoter	+	Throughout	Normal
HdhQ150	Mouse	150	C57BL/6JCBA	Full-length	Mouse htt promoter	+	Throughout	Normal
zQ175	Mouse	175	C57BL/6	Full-length	Mouse htt promoter	+	Striatum Cortex	Shorter lifespan

anxiety- and depressive-like phenotypes, have also been reported at three to four months of age, appearing before motor deficits (161,162). Both homozygous and heterozygous HdhQ111 display nuclear htt immunoreactivity in striatal MSNs at 1.5 and 5 months of age, respectively. However, reports of striatal atrophy have not been documented (157,163). Additionally, partial degeneration of the striatum, along with reactive gliosis, is detectable at around 24 months of age (157,163). Reduced body weight has been reported in this model (89). Table 6.2 summarizes some of the main features of each of the KI models.

6.3.2.3.2 CAG140

The CAG140 mouse model (also known as Htt140), generated by Scott Zeitlin's group, expresses 140 CAG repeats and represents a more robust phenotype than HdhQ111 mice (164) (Table 6.2). This KI model demonstrates motor dysfunction, such as reduced motor activity in the open-field test, gait anomalies, deficits in rotarod performance, and an initial increase in rearing behavior that declines with age (164). Moreover, they reveal long-term recognition memory, impairments, anxiety-like phenotypes, and decreased body weight (161,165). In the CAG140 model, the presence of NIIs has been demonstrated in several brain regions, including the striatum, cortex, nucleus accumbens, and olfactory tubercle (164). Furthermore, gliosis and atrophy have been observed in the striatum at 23 months of age (165,166). According to electrophysiological data, CAG140 mice demonstrate cortico-striatal circuitry dysfunction and striatal synaptic transmission impairments (167).

6.3.2.3.3 HdhQ150

In comparison to the KI models described above, the most severe pathology belongs to HdhQ150 mice (also known as CHL2). The HdhQ150 model was developed by Detloff and colleagues using the insertion of ~150 CAG repeats into exon 1 of the mouse htt homologue (168–170) (Table 6.2). At birth, these KI mice relatively display a normal phenotype. Starting at 40 weeks, this model reveals behavioral deficits, including disturbed rotarod performance, hypoactivity, clasping, and gait abnormalities that progress over time (84). Homozygous HdhQ150 mice demonstrate spatial learning deficits measured by the MWM test (171). In addition, striatal nuclear aggregates and significantly increased gliosis in the striatum have been reported (108,126). HdhQ150 mice show weight loss at 12 months (158,172), altered gene expression at 15 months (173), striatal axonal degeneration at 14 months (108), and striatal cell loss at approximately two years of age (165,168). Structural magnetic resonance imaging (MRI) techniques revealed that HdhQ150 animals undergo a progressive reduction in the striatal and cortical volume (172). Furthermore, the HdhQ150 model preferentially represents striatal NIIs, starting at around ten months of age (174). In addition to the striatum, NIIs are also present in the cortex. At 100 weeks of age, the evaluation of striatal dopamine, receptors (D1, and D2) binding sites exhibits a meaningful reduction in these binding sites (168,175).

6.3.2.3.4 zQ175

In addition to the KI models mentioned above, researchers have developed another KI mouse model called zQ175. This model originated as a spontaneous CAG expansion of the CAG140 mouse. The increased number of polyQ tracts in zQ175 leads to more prominent signs and an earlier disease onset (155). As early as one to two months of age, this model indicates motor impairments. At six months of age, zQ175 mice demonstrate transcriptional dysregulation (155). Moreover, they show cognitive deficits, hypoactivity, weight loss, cortical thinning, and decreased striatal dopamine (176). Based on some electrophysiological data, the zQ175 model shows diminished glutamatergic transmission (176,177). In zQ175 mice, striatal and cortical atrophy, along with the presence of NIIs in these brain regions, has been reported (155,176,178). These nuclear inclusions subsequently spread to other brain regions, including the nucleus accumbens, amygdala, hippocampus, hypothalamus, and thalamus (176). Compared to other described KI models, zQ175 mice exhibit a shorter lifespan (179) (Table 6.2).

6.3.2.4 Advantages and Pitfalls of KI-Based Models

In certain respects, KI mice serve as the most accurate models of htt protein pathology. They are genetically engineered to carry mutations in the htt gene, thereby mimicking the genetic basis of the disease as it occurs in humans. These models can replicate the specific mutations found in human patients, including the CAG repeat expansion. This specificity allows for a more accurate representation of the disease. By introducing human mutations into the mouse genome, these models better mimic the molecular and cellular mechanisms of HD, including protein aggregation and associated behavioral phenotypes. The stable genetic modification allows researchers to study the progression of HD over time within the same individuals, providing insights into the disease's trajectory and response to potential therapeutics, while also being suitable for studying late-onset HD. Many KI models exhibit behavioral changes consistent with the human condition, such as motor deficits and cognitive impairments, enabling researchers to test interventions and observe outcomes analogous to human symptoms.

While KI models provide valuable insights, they are still based on rodent physiology, which may not completely reflect human disease pathology and progression. Differences in brain structure and function can limit the translatability of findings. Consequently, these models may not capture the full spectrum of the disease's phenotypes. Furthermore, compared to transgenic models, their behavioral deficits are less prominent and require more time to develop. Some KI models may not sufficiently replicate later-stage symptoms or advanced pathology observed in human patients, potentially limiting their usefulness in studying disease progression and intervention efficacy.

6.4 CONCLUSION

The HD rodent models described earlier are among the most widely used in HD research. They have been pivotal in increasing our understanding of HD and advancing research into potential interventions. However, it is crucial to recognize that no single rodent model captures all aspects of human HD. Nevertheless, these models remain essential for preclinical studies, particularly in assessing new therapeutic approaches aimed at alleviating HD-related symptoms. Additionally, rodent models typically have shorter lifespans than humans, complicating the investigation of late-onset features of HD that may occur in older individuals. Therefore, there is a significant need to develop more diverse and accurate animal models in HD research to achieve a more comprehensive understanding of the condition.

REFERENCES

1. Zuccato C, Cattaneo E. Huntington's disease. Handb Exp Pharmacol [Internet]. 2014 [cited 2025 Apr 6]; 220:357–409. Available from: https://pubmed.ncbi.nlm.nih.gov/24668480/
2. Irfan Z, Khanam S, Karmakar V, Firdous SM, El Khier BSIA, Khan I, et al. Pathogenesis of Huntington's disease: An emphasis on molecular pathways and prevention by natural remedies. Brain Sci [Internet]. 2022 Oct 1 [cited 2025 Apr 6];12(10):1389. Available from: https://pmc.ncbi.nlm.nih.gov/articles/PMC9599635/
3. Karlovich CA, John RM, Ramirez L, Stainier DYR, Myers RM. Characterization of the Huntington's disease (HD) gene homologue in the zebrafish *Danio rerio*. Gene [Internet]. 1998 Sep 14 [cited 2025 Apr 6];217(1–2):117–25. Available from: https://pubmed.ncbi.nlm.nih.gov/9795179/
4. Tello JA, Williams HE, Eppler RM, Steinhilb ML, Khanna M. Animal models of neurodegenerative disease: Recent advances in fly highlight innovative approaches to drug discovery. Front Mol Neurosci [Internet]. 2022 Apr 19 [cited 2025 Apr 6];15. Available from: https://pubmed.ncbi.nlm.nih.gov/35514431/
5. Yin P, Li S, Li XJ, Yang W. New pathogenic insights from large animal models of neurodegenerative diseases. Protein Cell [Internet]. 2022 Oct 1 [cited 2025 Apr 6];13(10):707–20. Available from: https://pubmed.ncbi.nlm.nih.gov/35334073/
6. Ramaswamy S, McBride JL, Kordower JH. Animal models of Huntington's disease. ILAR J [Internet]. 2007 [cited 2025 Apr 6];48(4):356–73. Available from: https://pubmed.ncbi.nlm.nih.gov/17712222/
7. Guyot MC, Hantraye P, Dolan R, Palfi S, Maziére M, Brouillet E. Quantifiable bradykinesia, gait abnormalities and Huntington's disease- like striatal lesions in rats chronically treated with 3-nitropropionic acid. Neuroscience [Internet]. 1997 Apr 28 [cited 2025 Apr 6];79(1):45–56. Available from: https://pubmed.ncbi.nlm.nih.gov/9178864/
8. Tattersfield AS, Croon RJ, Liu YW, Kells AP, Faull RLM, Connor B. Neurogenesis in the striatum of the quinolinic acid lesion model of Huntington's disease. Neuroscience [Internet]. 2004 [cited 2025 Apr 6];127(2):319–32. Available from: https://pubmed.ncbi.nlm.nih.gov/15262322/
9. Beal MF, Ferrante RJ, Swartz KJ, Kowall NW. Chronic quinolinic acid lesions in rats closely resemble Huntington's disease. J Neurosci [Internet]. 1991 [cited 2025 Apr 6];11(6):1649–59. Available from: https://pubmed.ncbi.nlm.nih.gov/1710657/
10. MacDonald ME, Ambrose CM, Duyao MP, Myers RH, Lin C, Srinidhi L, et al. A novel gene containing a trinucleotide repeat that is expanded and unstable on Huntington's disease chromosomes. Cell. 1993 Mar 26;72(6):971–83.
11. Sharp AH, Loev SJ, Schilling G, Li SH, Li XJ, Bao J, et al. Widespread expression of Huntington's disease gene (IT15) protein product. Neuron [Internet]. 1995 [cited 2025 Apr 6];14(5):1065–74. Available from: https://pubmed.ncbi.nlm.nih.gov/7748554/
12. DiFiglia M, Sapp E, Chase KO, Davies SW, Bates GP, Vonsattel JP, et al. Aggregation of huntingtin in neuronal intranuclear inclusions and dystrophic neurites in brain. Science [Internet]. 1997 Sep 26 [cited 2025 Apr 6];277(5334):1990–3. Available from: https://pubmed.ncbi.nlm.nih.gov/9302293/
13. Becher MW, Kotzuk JA, Sharp AH, Davies SW, Bates GP, Price DL, et al. Intranuclear neuronal inclusions in Huntington's disease and dentatorubral and pallidoluysian atrophy: correlation between the density of inclusions and IT15 CAG triplet repeat length. Neurobiol Dis [Internet]. 1998 [cited 2025 Apr 6];4(6):387–97. Available from: https://pubmed.ncbi.nlm.nih.gov/9666478/
14. Maat-Schieman MLC, Dorsman JC, Smoor MA, Siesling S, Van Duinen SG, Verschuuren JJGM, et al. Distribution of inclusions in neuronal nuclei and dystrophic neurites in Huntington disease brain. J Neuropathol Exp Neurol [Internet]. 1999 [cited 2025 Apr 6];58(2):129–37. Available from: https://pubmed.ncbi.nlm.nih.gov/10029096/
15. Reiner A, Del Mar N, Deng YP, Meade CA, Sun Z, Goldowitz D. R6/2 neurons with intranuclear inclusions survive for prolonged periods in the brains of chimeric mice. J Comp Neurol [Internet]. 2007 Dec 20 [cited 2025 Apr 6];505(6):603–29. Available from: https://onlinelibrary.wiley.com/doi/full/10.1002/cne.21515
16. Arrasate M, Mitra S, Schweitzer ES, Segal MR, Finkbeiner S. Inclusion body formation reduces levels of mutant huntingtin and the risk of neuronal death. Nature [Internet]. 2004 Oct 14 [cited 2025 Apr 6];431(7010):805–10. Available from: https://pubmed.ncbi.nlm.nih.gov/15483602/
17. Ho LW, Carmichael J, Swartz J, Wyttenbach A, Rankin J, Rubinsztein DC. The molecular biology of Huntington's disease. Psychol Med [Internet]. 2001 [cited 2025 Apr 6];31(1):3–14. Available from: https://pubmed.ncbi.nlm.nih.gov/11200958/

18. Duyao M, Ambrose C, Myers R, Novelletto A, Persichetti F, Frontali M, et al. Trinucleotide repeat length instability and age of onset in Huntington's disease. Nat Genet [Internet]. 1993 [cited 2025 Apr 6]; 4(4):387–92. Available from: https://pubmed.ncbi.nlm.nih.gov/8401587/
19. Swami M, Hendricks AE, Gillis T, Massood T, Mysore J, Myers RH, et al. Somatic expansion of the Huntington's disease CAG repeat in the brain is associated with an earlier age of disease onset. Hum Mol Genet [Internet]. 2009 [cited 2025 Apr 6];18(16):3039–47. Available from: https://pubmed.ncbi.nlm.nih.gov/19465745/
20. Lee JM, Correia K, Loupe J, Kim KH, Barker D, Hong EP, et al. CAG repeat not polyglutamine length determines timing of Huntington's disease onset. Cell [Internet]. 2019 Aug 8 [cited 2025 Apr 6]; 178(4):887–900.e14. Available from: https://pubmed.ncbi.nlm.nih.gov/31398342/
21. Bates GP, Dorsey R, Gusella JF, Hayden MR, Kay C, Leavitt BR, et al. Huntington disease. Nat Rev Dis Primers [Internet]. 2015 Apr 23 [cited 2025 Apr 6];1. Available from: https://pubmed.ncbi.nlm.nih.gov/27188817/
22. Semaka A, Kay C, Doty C, Collins JA, Bijlsma EK, Richards F, et al. CAG size-specific risk estimates for intermediate allele repeat instability in Huntington disease. J Med Genet [Internet]. 2013 [cited 2025 Apr 6];50(10):696–703. Available from: https://pubmed.ncbi.nlm.nih.gov/23896435/
23. Lee JM, Wheeler VC, Chao MJ, Vonsattel JPG, Pinto RM, Lucente D, et al. Identification of genetic factors that modify clinical onset of Huntington's disease. Cell [Internet]. 2015 Aug 1 [cited 2025 Apr 6];162(3):516–26. Available from: https://pubmed.ncbi.nlm.nih.gov/26232222/
24. Becanovic K, Nørremølle A, Neal SJ, Kay C, Collins JA, Arenillas D, et al. A SNP in the HTT promoter alters NF-κB binding and is a bidirectional genetic modifier of Huntington disease. Nat Neurosci [Internet]. 2015 Jun 28 [cited 2025 Apr 6];18(6):807–16. Available from: https://pubmed.ncbi.nlm.nih.gov/25938884/
25. Rüb U, Hentschel M, Stratmann K, Brunt E, Heinsen H, Seidel K, et al. Huntington's disease (HD): Degeneration of select nuclei, widespread occurrence of neuronal nuclear and axonal inclusions in the brainstem. Brain Pathol [Internet]. 2014 [cited 2025 Apr 6];24(3):247. Available from: https://pmc.ncbi.nlm.nih.gov/articles/PMC4160739/
26. Rosas HD, Liu AK, Hersch S, Glessner M, Ferrante RJ, Salat DH, et al. Regional and progressive thinning of the cortical ribbon in Huntington's disease. Neurology [Internet]. 2002 Mar 12 [cited 2025 Apr 6];58(5):695–701. Available from: https://pubmed.ncbi.nlm.nih.gov/11889230/
27. De La Monte SM, Vonsattel JP, Richardson EP. Morphometric demonstration of atrophic changes in the cerebral cortex, white matter, and neostriatum in Huntington's disease. J Neuropathol Exp Neurol [Internet]. 1988 [cited 2025 Apr 6];47(5):516–25. Available from: https://pubmed.ncbi.nlm.nih.gov/2971785/
28. Harjes P, Wanker EE. The hunt for huntingtin function: Interaction partners tell many different stories. Trends Biochem Sci [Internet]. 2003 Aug 1 [cited 2025 Apr 6];28(8):425–33. Available from: https://pubmed.ncbi.nlm.nih.gov/12932731/
29. Ochaba J, Lukacsovich T, Csikos G, Zheng S, Margulis J, Salazar L, et al. Potential function for the huntingtin protein as a scaffold for selective autophagy. Proc Natl Acad Sci USA [Internet]. 2014 Nov 25 [cited 2025 Apr 6];111(47):16889–94. Available from: https://pubmed.ncbi.nlm.nih.gov/25385587/
30. Caviston JP, Holzbaur ELF. Huntingtin as an essential integrator of intracellular vesicular trafficking. Trends Cell Biol [Internet]. 2009 Apr [cited 2025 Apr 6];19(4):147–55. Available from: https://pubmed.ncbi.nlm.nih.gov/19269181/
31. Browne SE, Beal MF. The energetics of Huntington's disease. Neurochem Res [Internet]. 2004 Mar [cited 2025 Apr 6];29(3):531–46. Available from: https://pubmed.ncbi.nlm.nih.gov/15038601/
32. Mochel F, Durant B, Meng X, O'Callaghan J, Yu H, Brouillet E, et al. Early alterations of brain cellular energy homeostasis in Huntington disease models. J Biol Chem [Internet]. 2012 Jan 6 [cited 2025 Apr 6];287(2):1361–70. Available from: https://pubmed.ncbi.nlm.nih.gov/22123819/
33. Miller JRC, Lo KK, Andre R, Hensman Moss DJ, Träger U, Stone TC, et al. RNA-Seq of Huntington's disease patient myeloid cells reveals innate transcriptional dysregulation associated with proinflammatory pathway activation. Hum Mol Genet [Internet]. 2016 [cited 2025 Apr 6];25(14):2893–904. Available from: https://pubmed.ncbi.nlm.nih.gov/27170315/
34. Ehrlich ME. Huntington's disease and the striatal medium spiny neuron: Cell-autonomous and non-cell-autonomous mechanisms of disease. Neurotherapeutics [Internet]. 2012 Apr [cited 2025 Apr 6]; 9(2):270–84. Available from: https://pubmed.ncbi.nlm.nih.gov/22441874/
35. Johnson EB, Ziegler G, Penny W, Rees G, Tabrizi SJ, Scahill RI, et al. Dynamics of cortical degeneration over a decade in Huntington's disease. Biol Psychiatry [Internet]. 2021 Apr 15 [cited 2025 Apr 6]; 89(8):807–16. Available from: https://pubmed.ncbi.nlm.nih.gov/33500176/

36. Jankovic J, Roos RAC. Chorea associated with Huntington's disease: To treat or not to treat? Mov Disord [Internet]. 2014 Sep 15 [cited 2025 Apr 6];29(11):1414–8. Available from: https://pubmed.ncbi.nlm.nih.gov/25156927/
37. Roos RAC. Huntington's disease: A clinical review. Orphanet J Rare Dis [Internet]. 2010 [cited 2025 Apr 6];5(1). Available from: https://pubmed.ncbi.nlm.nih.gov/21171977/
38. Snowden JS, Craufurd D, Thompson J, Neary D. Psychomotor, executive, and memory function in preclinical Huntington's disease. J Clin Exp Neuropsychol [Internet]. 2002 [cited 2025 Apr 6];24(2): 133–45. Available from: https://pubmed.ncbi.nlm.nih.gov/11992196/
39. Paulsen JS. Cognitive impairment in Huntington disease: Diagnosis and treatment. Curr Neurol Neurosci Rep [Internet]. 2011 Oct [cited 2025 Apr 6];11(5):474–83. Available from: https://pubmed.ncbi.nlm.nih.gov/21861097/
40. Julien CL, Thompson JC, Wild S, Yardumian P, Snowden JS, Turner G, et al. Psychiatric disorders in preclinical Huntington's disease. J Neurol Neurosurg Psychiatry [Internet]. 2007 Sep [cited 2025 Apr 6]; 78(9):939–43. Available from: https://pubmed.ncbi.nlm.nih.gov/17178819/
41. Pillon B, Dubois B, Ploska A, Agid Y. Severity and specificity of cognitive impairment in Alzheimer's, Huntington's, and Parkinson's diseases and progressive supranuclear palsy. Neurology [Internet]. 1991 [cited 2025 Apr 6];41(5):634–43. Available from: https://pubmed.ncbi.nlm.nih.gov/1827513/
42. Butters N, Wolfe J, Martone M, Granholm E, Cermak LS. Memory disorders associated with Huntington's disease: verbal recall, verbal recognition and procedural memory. Neuropsychologia [Internet]. 1985 [cited 2025 Apr 6];23(6):729–43. Available from: https://pubmed.ncbi.nlm.nih.gov/2934642/
43. Van Duijn E, Craufurd D, Hubers AAM, Giltay EJ, Bonelli R, Rickards H, et al. Neuropsychiatric symptoms in a European Huntington's disease cohort (REGISTRY). J Neurol Neurosurg Psychiatry [Internet]. 2014 Dec 1 [cited 2025 Apr 6];85(12):1411–8. Available from: https://pubmed.ncbi.nlm.nih.gov/24828898/
44. Galts CPC, Bettio LEB, Jewett DC, Yang CC, Brocardo PS, Rodrigues ALS, et al. Depression in neurodegenerative diseases: Common mechanisms and current treatment options. Neurosci Biobehav Rev [Internet]. 2019 Jul 1 [cited 2025 Apr 6];102:56–84. Available from: https://pubmed.ncbi.nlm.nih.gov/30995512/
45. Sanberg PR, Fibiger HC, Mark RF. Body weight and dietary factors in Huntington's disease patients compared with matched controls. Med J Aust [Internet]. 1981 [cited 2025 Apr 6];1(8):407–9. Available from: https://pubmed.ncbi.nlm.nih.gov/6454826/
46. Morton AJ. Sleep and circadian rhythm dysfunction in animal models of Huntington's disease. J Huntingtons Dis [Internet]. 2023 [cited 2025 Apr 6];12(2):133–48. Available from: https://pubmed.ncbi.nlm.nih.gov/37334613/
47. Koch ET, Sepers MD, Cheng J, Raymond LA. Early changes in striatal activity and motor kinematics in a Huntington's disease mouse model. Mov Disord. 2022 Oct;37(10):2021–32.
48. Kim A, Lalonde K, Truesdell A, Gomes Welter P, Brocardo PS, Rosenstock TR, et al. New avenues for the treatment of Huntington's disease. Int J Mol Sci. 2021 Aug;22(16).
49. Landwehrmeyer GB, Standaert DG, Testa CM, Penney JBJ, Young AB. NMDA receptor subunit mRNA expression by projection neurons and interneurons in rat striatum. J Neurosci. 1995 Jul;15(7 Pt 2): 5297–307.
50. R. Rosenstock T, I. Duarte A, Cristina Rego A. Mitochondrial-associated metabolic changes and neurodegeneration in Huntington's disease – From clinical features to the bench. Curr Drug Targets. 2012 Nov 11;11(10):1218–36.
51. Estrada Sánchez AM, Mejía-Toiber J, Massieu L. Excitotoxic neuronal death and the pathogenesis of Huntington's disease. Arch Med Res [Internet]. 2008 Apr [cited 2025 Apr 7];39(3):265–76. Available from: https://pubmed.ncbi.nlm.nih.gov/18279698/
52. Coyle JT, Schwarcz R. Lesion of striatal neurones with kainic acid provides a model for Huntington's chorea. Nature [Internet]. 1976 [cited 2025 Apr 7];263(5574):244–6. Available from: https://pubmed.ncbi.nlm.nih.gov/8731/
53. Schwarcz R, Coyle JT. Striatal lesions with kainic acid: Neurochemical characteristics. Brain Res [Internet]. 1977 May 27 [cited 2025 Apr 7];127(2):235–49. Available from: https://pubmed.ncbi.nlm.nih.gov/16686/
54. Rana N, Kapil L, Singh C, Singh A. Modeling Huntington's disease: An insight on in-vitro and in-vivo models. Behav Brain Res [Internet]. 2024 Feb 29 [cited 2025 Apr 7];459. Available from: https://pubmed.ncbi.nlm.nih.gov/37952684/

55. Schwarcz R, Guidetti P, Sathyasaikumar K V., Muchowski PJ. Of mice, rats and men: Revisiting the quinolinic acid hypothesis of Huntington's disease. Prog Neurobiol [Internet]. 2010 Feb 9 [cited 2025 Apr 7];90(2):230–45. Available from: https://pubmed.ncbi.nlm.nih.gov/19394403/
56. Schwarcz R, Okuno E, White RJ, Bird ED, Whetsell WO. 3-Hydroxyanthranilate oxygenase activity is increased in the brains of Huntington disease victims. Proc Natl Acad Sci USA [Internet]. 1988 [cited 2025 Apr 7];85(11):4079. Available from: https://pmc.ncbi.nlm.nih.gov/articles/PMC280365/
57. Guidetti P, Luthi-Carter RE, Augood SJ, Schwarcz R. Neostriatal and cortical quinolinate levels are increased in early grade Huntington's disease. Neurobiol Dis [Internet]. 2004 Dec [cited 2025 Apr 7]; 17(3):455–61. Available from: https://pubmed.ncbi.nlm.nih.gov/15571981/
58. McLin JP, Thompson LM, Steward O. Differential susceptibility to striatal neurodegeneration induced by quinolinic acid and kainate in inbred, outbred and hybrid mouse strains. Eur J Neurosci [Internet]. 2006 Dec [cited 2025 Apr 7];24(11):3134–40. Available from: https://pubmed.ncbi.nlm.nih.gov/17156374/
59. Ribeiro CAJ, Grando V, Dutra Filho CS, Wannmacher CMD, Wajner M. Evidence that quinolinic acid severely impairs energy metabolism through activation of NMDA receptors in striatum from developing rats. J Neurochem [Internet]. 2006 Dec [cited 2025 Apr 7];99(6):1531–42. Available from: https://pubmed.ncbi.nlm.nih.gov/17230642/
60. Vazey EM, Chen K, Hughes SM, Connor B. Transplanted adult neural progenitor cells survive, differentiate and reduce motor function impairment in a rodent model of Huntington's disease. Exp Neurol [Internet]. 2006 Jun [cited 2025 Apr 7];199(2):384–96. Available from: https://pubmed.ncbi.nlm.nih.gov/16626705/
61. McBride JL, Behrstock SP, Chen EY, Jakel RJ, Siegel I, Svendsen CN, et al. Human neural stem cell transplants improve motor function in a rat model of Huntington's disease. J Comp Neurol [Internet]. 2004 Jul 19 [cited 2025 Apr 7];475(2):211–9. Available from: https://pubmed.ncbi.nlm.nih.gov/15211462/
62. Isacson O, Brundin P, Kelly PAT, Gage FH, Björklund A. Functional neuronal replacement by grafted striatal neurones in the ibotenic acid-lesioned rat striatum. Nature [Internet]. 1984 [cited 2025 Apr 7]; 311(5985):458–60. Available from: https://pubmed.ncbi.nlm.nih.gov/6482962/
63. Furtado JCS, Mazurek MF. Behavioral characterization of quinolinate-induced lesions of the medial striatum: Relevance for Huntington's disease. Exp Neurol [Internet]. 1996 [cited 2025 Apr 7];138(1): 158–68. Available from: https://pubmed.ncbi.nlm.nih.gov/8593891/
64. Shear DA, Dong J, Haik-Creguer KL, Bazzett TJ, Albin RL, Dunbar GL. Chronic administration of quinolinic acid in the rat striatum causes spatial learning deficits in a radial arm water maze task. Exp Neurol [Internet]. 1998 [cited 2025 Apr 7];150(2):305–11. Available from: https://pubmed.ncbi.nlm.nih.gov/9527900/
65. Damiano M, Galvan L, Déglon N, Brouillet E. Mitochondria in Huntington's disease. Biochim Biophys Acta [Internet]. 2010 Jan [cited 2025 Apr 7];1802(1):52–61. Available from: https://pubmed.ncbi.nlm.nih.gov/19682570/
66. Browne SE, Bowling AC, MacGarvey U, Baik MJ, Berger SC, Muqit MMK, et al. Oxidative damage and metabolic dysfunction in Huntington's disease: Selective vulnerability of the basal ganglia. Ann Neurol [Internet]. 1997 May [cited 2025 Apr 7];41(5):646–53. Available from: https://pubmed.ncbi.nlm.nih.gov/9153527/
67. Gu M, Gash MT, Mann VM, Javoy-Agid F, Cooper JM, Schapira AHV. Mitochondrial defect in Huntington's disease caudate nucleus. Ann Neurol [Internet]. 1996 [cited 2025 Apr 7];39(3):385–9. Available from: https://pubmed.ncbi.nlm.nih.gov/8602759/
68. Kim J, Moody JP, Edgerly CK, Bordiuk OL, Cormier K, Smith K, et al. Mitochondrial loss, dysfunction and altered dynamics in Huntington's disease. Hum Mol Genet [Internet]. 2010 Jul 21 [cited 2025 Apr 7];19(20):3919–35. Available from: https://pubmed.ncbi.nlm.nih.gov/20660112/
69. Ludolph AC, he F, Spencer PS, Hammerstad J, Sabri M. 3-Nitropropionic acid-exogenous animal neurotoxin and possible human striatal toxin. Can J Neurol Sci [Internet]. 1991 [cited 2025 Apr 7];18(4): 492–8. Available from: https://pubmed.ncbi.nlm.nih.gov/1782616/
70. Suganya SN, Sumathi T. Effect of rutin against a mitochondrial toxin, 3-nitropropionicacid induced biochemical, behavioral and histological alterations-a pilot study on Huntington's disease model in rats. Metab Brain Dis [Internet]. 2017 Apr 1 [cited 2025 Apr 7];32(2):471–81. Available from: https://pubmed.ncbi.nlm.nih.gov/27928694/
71. Yuan C, Zheng L, Zhao Y. Protective effect of 3-n-butylphthalide against intrastriatal injection of malonic acid-induced neurotoxicity and biochemical alteration in rats. Biomed Pharmacother [Internet]. 2022 Nov 1 [cited 2025 Apr 7];155. Available from: https://pubmed.ncbi.nlm.nih.gov/36095961/
72. Coles CJ, Edmondson DE, Singer TP. Inactivation of succinate dehydrogenase by 3-nitropropionate. J Biol Chem. 1979 Jun 25;254(12):5161–7.

73. Ouary S, Bizat N, Altairac S, Ménétrat H, Mittoux V, Condé F, et al. Major strain differences in response to chronic systemic administration of the mitochondrial toxin 3-nitropropionic acid in rats: implications for neuroprotection studies. Neuroscience [Internet]. 2000 May [cited 2025 Apr 7];97(3):521–30. Available from: https://pubmed.ncbi.nlm.nih.gov/10828534/
74. Borlongan C V., Koutouzis TK, Freeman TB, Hauser RA, Cahill DW, Sanberg PR. Hyperactivity and hypoactivity in a rat model of Huntington's disease: the systemic 3 nitropropionic acid model. Brain Res Brain Res Protoc [Internet]. 1997 Aug [cited 2025 Apr 7];1(3):253–7. Available from: https://pubmed.ncbi.nlm.nih.gov/9385062/
75. Nittari G, Roy P, Martinelli I, Bellitto V, Tomassoni D, Traini E, et al. Rodent models of Huntington's disease: An overview. Biomedicines [Internet]. 2023 Dec 1 [cited 2025 Apr 7];11(12). Available from: https://pubmed.ncbi.nlm.nih.gov/38137552/
76. Mangiarini L, Sathasivam K, Seller M, Cozens B, Harper A, Hetherington C, et al. Exon 1 of the HD gene with an expanded CAG repeat is sufficient to cause a progressive neurological phenotype in transgenic mice. Cell [Internet]. 1996 Nov 1 [cited 2025 Apr 7];87(3):493–506. Available from: https://pubmed.ncbi.nlm.nih.gov/8898202/
77. Davies SW, Turmaine M, Cozens BA, DiFiglia M, Sharp AH, Ross CA, et al. Formation of neuronal intranuclear inclusions underlies the neurological dysfunction in mice transgenic for the HD mutation. Cell [Internet]. 1997 Aug 8 [cited 2025 Apr 7];90(3):537–48. Available from: https://pubmed.ncbi.nlm.nih.gov/9267033/
78. Lione LA, Carter RJ, Hunt MJ, Bates GP, Morton AJ, Dunnett SB. Selective discrimination learning impairments in mice expressing the human Huntington's disease mutation. J Neurosci [Internet]. 1999 Dec 1 [cited 2025 Apr 7];19(23):10428–37. Available from: https://pubmed.ncbi.nlm.nih.gov/10575040/
79. Hodges A, Hughes G, Brooks S, Elliston L, Holmans P, Dunnett SB, et al. Brain gene expression correlates with changes in behavior in the R6/1 mouse model of Huntington's disease. Genes Brain Behav [Internet]. 2008 Apr [cited 2025 Apr 7];7(3):288–99. Available from: https://pubmed.ncbi.nlm.nih.gov/17696994/
80. Pang TYC, Du X, Zajac MS, Howard ML, Hannan AJ. Altered serotonin receptor expression is associated with depression-related behavior in the R6/1 transgenic mouse model of Huntington's disease. Hum Mol Genet [Internet]. 2009 [cited 2025 Apr 7];18(4):753–66. Available from: https://pubmed.ncbi.nlm.nih.gov/19008301/
81. Pla P, Orvoen S, Saudou F, David DJ, Humbert S. Mood disorders in Huntington's disease: From behavior to cellular and molecular mechanisms. Front Behav Neurosci [Internet]. 2014 Apr 23 [cited 2025 Apr 7];8(APR). Available from: https://pubmed.ncbi.nlm.nih.gov/24795586/
82. Naver B, Stub C, Møller M, Fenger K, Hansen AK, Hasholt L, et al. Molecular and behavioral analysis of the R6/1 Huntington's disease transgenic mouse. Neuroscience [Internet]. 2003 [cited 2025 Apr 7];122(4):1049–57. Available from: https://pubmed.ncbi.nlm.nih.gov/14643771/
83. Milnerwood AJ, Cummings DM, Dallérac GM, Brown JY, Vatsavayai SC, Hirst MC, et al. Early development of aberrant synaptic plasticity in a mouse model of Huntington's disease. Hum Mol Genet [Internet]. 2006 May 15 [cited 2025 Apr 7];15(10):1690–703. Available from: https://pubmed.ncbi.nlm.nih.gov/16600988/
84. Farshim PP, Bates GP. Mouse models of Huntington's disease. Methods Mol Biol [Internet]. 2018 [cited 2025 Apr 7];1780:97–120. Available from: https://pubmed.ncbi.nlm.nih.gov/29856016/
85. Stack EC, Kubilus JK, Smith K, Cormier K, Del Signore SJ, Guelin E, et al. Chronology of behavioral symptoms and neuropathological sequela in R6/2 Huntington's disease transgenic mice. J Comp Neurol [Internet]. 2005 Oct 3 [cited 2025 Apr 7];490(4):354–70. Available from: https://pubmed.ncbi.nlm.nih.gov/16127709/
86. Hickey MA, Gallant K, Gross GG, Levine MS, Chesselet MF. Early behavioral deficits in R6/2 mice suitable for use in preclinical drug testing. Neurobiol Dis [Internet]. 2005 Oct [cited 2025 Apr 7];20(1):1–11. Available from: https://pubmed.ncbi.nlm.nih.gov/16137562/
87. Dodds L, Chen J, Berggren K, Fox J. Characterization of striatal neuronal loss and atrophy in the R6/2 mouse model of Huntington's disease. PLoS Curr [Internet]. 2014 [cited 2025 Apr 7];6(JAN): ecurrents.hd.48727b68b39b82d5fe350f753984bcf9. Available from: https://pmc.ncbi.nlm.nih.gov/articles/PMC3882322/
88. Cayzac S, Delcasso S, Paz V, Jeantet Y, Cho YH. Changes in striatal procedural memory coding correlate with learning deficits in a mouse model of Huntington disease. Proc Natl Acad Sci USA [Internet]. 2011 May 31 [cited 2025 Apr 7];108(22):9280–5. Available from: https://pubmed.ncbi.nlm.nih.gov/21576479/

89. Menalled L, El-Khodor BF, Patry M, Suárez-Fariñas M, Orenstein SJ, Zahasky B, et al. Systematic behavioral evaluation of Huntington's disease transgenic and knock-in mouse models. Neurobiol Dis [Internet]. 2009 Sep [cited 2025 Apr 7];35(3):319–36. Available from: https://pubmed.ncbi.nlm.nih.gov/19464370/
90. Chiu CT, Liu G, Leeds P, Chuang DM. Combined treatment with the mood stabilizers lithium and valproate produces multiple beneficial effects in transgenic mouse models of Huntington's disease. Neuropsychopharmacology [Internet]. 2011 Nov [cited 2025 Apr 7];36(12):2406–21. Available from: https://pubmed.ncbi.nlm.nih.gov/21796107/
91. Li H, Li SH, Yu ZX, Shelbourne P, Li XJ. Huntingtin aggregate-associated axonal degeneration is an early pathological event in Huntington's disease mice. J Neurosci [Internet]. 2001 Nov 1 [cited 2025 Apr 7];21(21):8473–81. Available from: https://pubmed.ncbi.nlm.nih.gov/11606636/
92. Morton AJ, Lagan MA, Skepper JN, Dunnett SB. Progressive formation of inclusions in the striatum and hippocampus of mice transgenic for the human Huntington's disease mutation. J Neurocytol [Internet]. 2000 Sep 1 [cited 2025 Apr 7];29(9):679–702. Available from: https://pubmed.ncbi.nlm.nih.gov/11353291/
93. Peterseén Å, Gil J, Maat-Schieman MLC, Björkqvist M, Tanila H, Araújo IM, et al. Orexin loss in Huntington's disease. Hum Mol Genet [Internet]. 2005 Jan 1 [cited 2025 Apr 7];14(1):39–47. Available from: https://pubmed.ncbi.nlm.nih.gov/15525658/
94. Ganjavi H, Shapiro CM. Hypocretin/Orexin: A molecular link between sleep, energy regulation, and pleasure. J Neuropsychiatry Clin Neurosci [Internet]. 2007 [cited 2025 Apr 7];19(4):413–9. Available from: https://pubmed.ncbi.nlm.nih.gov/18070844/
95. Luthi-Carter R, Strand A, Peters NL, Solano SM, Hollingsworth ZR, Menon AS, et al. Decreased expression of striatal signaling genes in a mouse model of Huntington's disease. Hum Mol Genet [Internet]. 2000 May 22 [cited 2025 Apr 7];9(9):1259–71. Available from: https://pubmed.ncbi.nlm.nih.gov/10814708/
96. Bibb JA, Yan Z, Svenningsson P, Snyder GL, Pieribone VA, Horiuchi A, et al. Severe deficiencies in dopamine signaling in presymptomatic Huntington's disease mice. Proc Natl Acad Sci USA [Internet]. 2000 Jun 6 [cited 2025 Apr 7];97(12):6809–14. Available from: https://pubmed.ncbi.nlm.nih.gov/10829080/
97. Mielcarek M, Inuabasi L, Bondulich MK, Muller T, Osborne GF, Franklin SA, et al. Dysfunction of the CNS-heart axis in mouse models of Huntington's disease. PLoS Genet [Internet]. 2014 Aug 7 [cited 2025 Apr 7];10(8). Available from: https://pubmed.ncbi.nlm.nih.gov/25101683/
98. Ribchester RR, Thomson D, Wood NI, Hinks T, Gillingwater TH, Wishart TM, et al. Progressive abnormalities in skeletal muscle and neuromuscular junctions of transgenic mice expressing the Huntington's disease mutation. Eur J Neurosci [Internet]. 2004 Dec [cited 2025 Apr 7];20(11):3092–114. Available from: https://pubmed.ncbi.nlm.nih.gov/15579164/
99. Bolivar VJ, Manley K, Messer A. Exploratory activity and fear conditioning abnormalities develop early in R6/2 Huntington's disease transgenic mice. Behav Neurosci [Internet]. 2003 Dec [cited 2025 Apr 7];117(6):1233–42. Available from: https://pubmed.ncbi.nlm.nih.gov/14674843/
100. Cha JHJ, Frey AS, Alsdorf SA, Kerner JA, Kosinski CM, Mangiarini L, et al. Altered neurotransmitter receptor expression in transgenic mouse models of Huntington's disease. Philos Trans R Soc Lond B Biol Sci [Internet]. 1999 Jun 29 [cited 2025 Apr 7];354(1386):981–9. Available from: https://pubmed.ncbi.nlm.nih.gov/10434296/
101. Murphy KPSJ, Carter RJ, Lione LA, Mangiarini L, Mahal A, Bates GP, et al. Abnormal synaptic plasticity and impaired spatial cognition in mice transgenic for exon 1 of the human Huntington's disease mutation. J Neurosci [Internet]. 2000 Jul 1 [cited 2025 Apr 7];20(13):5115–23. Available from: https://pubmed.ncbi.nlm.nih.gov/10864968/
102. Schilling G, Becher MW, Sharp AH, Jinnah HA, Duan K, Kotzuk JA, et al. Intranuclear inclusions and neuritic aggregates in transgenic mice expressing a mutant N-terminal fragment of huntingtin. Hum Mol Genet [Internet]. 1999 Mar 1 [cited 2025 Apr 7];8(3):397–407. Available from: https://pubmed.ncbi.nlm.nih.gov/9949199/
103. Rangel-Barajas C, Rebec GV. Overview of Huntington's disease models: Neuropathological, molecular, and behavioral differences. Curr Protoc Neurosci [Internet]. 2018 Apr 1 [cited 2025 Apr 7];83(1):e47. Available from: https://pubmed.ncbi.nlm.nih.gov/30040221/
104. Ferrante RJ. Mouse models of Huntington's disease and methodological considerations for therapeutic trials. Biochim Biophys Acta [Internet]. 2009 Jun [cited 2025 Apr 7];1792(6):506–20. Available from: https://pubmed.ncbi.nlm.nih.gov/19362590/

105. McBride JL, Ramaswamy S, Gasmi M, Bartus RT, Herzog CD, Brandon EP, et al. Viral delivery of glial cell line-derived neurotrophic factor improves behavior and protects striatal neurons in a mouse model of Huntington's disease. Proc Natl Acad Sci USA [Internet]. 2006 Jun 13 [cited 2025 Apr 7]; 103(24):9345–50. Available from: https://pubmed.ncbi.nlm.nih.gov/16751280/
106. Aggarwal M, Duan W, Hou Z, Rakesh N, Peng Q, Ross CA, et al. Spatiotemporal mapping of brain atrophy in mouse models of Huntington's disease using longitudinal in vivo magnetic resonance imaging. Neuroimage [Internet]. 2012 May 1 [cited 2025 Apr 7];60(4):2086–95. Available from: https://pubmed.ncbi.nlm.nih.gov/22342677/
107. Cheng Y, Peng Q, Hou Z, Aggarwal M, Zhang J, Mori S, et al. Structural MRI detects progressive regional brain atrophy and neuroprotective effects in N171-82Q Huntington's disease mouse model. Neuroimage [Internet]. 2011 Jun 1 [cited 2025 Apr 7];56(3):1027–34. Available from: https://pubmed.ncbi.nlm.nih.gov/21320608/
108. Yu ZX, Li SH, Evans J, Pillarisetti A, Li H, Li XJ. Mutant huntingtin causes context-dependent neurodegeneration in mice with Huntington's disease. J Neurosci [Internet]. 2003 Mar 15 [cited 2025 Apr 7];23(6):2193–202. Available from: https://pubmed.ncbi.nlm.nih.gov/12657678/
109. Li SH, Yu ZX, Li CL, Nguyen HP, Zhou YX, Deng C, et al. Lack of huntingtin -associated protein-1 causes neuronal death resembling hypothalamic degeneration in Huntington's disease. J Neurosci [Internet]. 2003 Jul 30 [cited 2025 Apr 7];23(17):6956–64. Available from: https://pubmed.ncbi.nlm.nih.gov/12890790/
110. Chan EYW, Luthi-Carter R, Strand A, Solano SM, Hanson SA, DeJohn MM, et al. Increased huntingtin protein length reduces the number of polyglutamine-induced gene expression changes in mouse models of Huntington's disease. Hum Mol Genet. 2002 Aug;11(17):1939–51.
111. Gray M, Shirasaki DI, Cepeda C, André VM, Wilburn B, Lu XH, et al. Full-length human mutant huntingtin with a stable polyglutamine repeat can elicit progressive and selective neuropathogenesis in BACHD mice. J Neurosci. 2008 Jun;28(24):6182–95.
112. Slow EJ, van Raamsdonk J, Rogers D, Coleman SH, Graham RK, Deng Y, et al. Selective striatal neuronal loss in a YAC128 mouse model of Huntington disease. Hum Mol Genet. 2003 Jul;12(13):1555–67.
113. Nittari G, Roy P, Martinelli I, Bellitto V, Tomassoni D, Traini E, et al. Rodent Models of Huntington's Disease: An Overview. Biomedicines. 2023 Dec;11(12).
114. Van Raamsdonk JM, Murphy Z, Slow EJ, Leavitt BR, Hayden MR. Selective degeneration and nuclear localization of mutant huntingtin in the YAC128 mouse model of Huntington disease. Hum Mol Genet. 2005 Dec;14(24):3823–35.
115. Wang N, Gray M, Lu XH, Cantle JP, Holley SM, Greiner E, et al. Neuronal targets for reducing mutant huntingtin expression to ameliorate disease in a mouse model of Huntington's disease. Nat Med. 2014 May;20(5):536–41.
116. Xie Y, Hayden MR, Xu B. BDNF overexpression in the forebrain rescues Huntington's disease phenotypes in YAC128 mice. J Neurosci. 2010 Nov;30(44):14708–18.
117. Teo RTY, Hong X, Yu-Taeger L, Huang Y, Tan LJ, Xie Y, et al. Structural and molecular myelination deficits occur prior to neuronal loss in the YAC128 and BACHD models of Huntington disease. Hum Mol Genet. 2016 Jul;25(13):2621–32.
118. Hodgson JG, Agopyan N, Gutekunst CA, Leavitt BR, LePiane F, Singaraja R, et al. A YAC mouse model for Huntington's disease with full-length mutant huntingtin, cytoplasmic toxicity, and selective striatal neurodegeneration. Neuron. 1999 May;23(1):181–92.
119. Benn CL, Slow EJ, Farrell LA, Graham R, Deng Y, Hayden MR, et al. Glutamate receptor abnormalities in the YAC128 transgenic mouse model of Huntington's disease. Neuroscience. 2007 Jun;147(2):354–72.
120. Van Raamsdonk JM, Pearson J, Slow EJ, Hossain SM, Leavitt BR, Hayden MR. Cognitive dysfunction precedes neuropathology and motor abnormalities in the YAC128 mouse model of Huntington's disease. J Neurosci. 2005 Apr;25(16):4169–80.
121. Pouladi MA, Graham RK, Karasinska JM, Xie Y, Santos RD, Petersén A, et al. Prevention of depressive behaviour in the YAC128 mouse model of Huntington disease by mutation at residue 586 of huntingtin. Brain. 2009 Apr;132(Pt 4):919–32.
122. Garcia-Miralles M, Ooi J, Ferrari Bardile C, Tan LJ, George M, Drum CL, et al. Treatment with the MAO-A inhibitor clorgyline elevates monoamine neurotransmitter levels and improves affective phenotypes in a mouse model of Huntington disease. Exp Neurol. 2016 Apr;278:4–10.
123. Brooks SP, Janghra N, Higgs G V, Bayram-Weston Z, Heuer A, Jones L, et al. Selective cognitive impairment in the YAC128 Huntington's disease mouse. Brain Res Bull. 2012 Jun;88(2–3):121–9.

124. Pouladi MA, Brillaud E, Xie Y, Conforti P, Graham RK, Ehrnhoefer DE, et al. NP03, a novel low-dose lithium formulation, is neuroprotective in the YAC128 mouse model of Huntington disease. Neurobiol Dis. 2012 Dec;48(3):282–9.
125. Björkqvist M, Petersén A, Nielsen J, Ecker D, Mulder H, Hayden MR, et al. Cerebrospinal fluid levels of orexin-A are not a clinically useful biomarker for Huntington disease, Clinical genetics. Denmark; 2006;70:78–79.
126. Bayram-Weston Z, Torres EM, Jones L, Dunnett SB, Brooks SP. Light and electron microscopic characterization of the evolution of cellular pathology in the Hdh(CAG)150 Huntington's disease knock-in mouse. Brain Res Bull. 2012 Jun;88(2–3):189–98.
127. Pouladi MA, Stanek LM, Xie Y, Franciosi S, Southwell AL, Deng Y, et al. Marked differences in neurochemistry and aggregates despite similar behavioural and neuropathological features of Huntington disease in the full-length BACHD and YAC128 mice. Hum Mol Genet. 2012 May;21(10):2219–32.
128. Van Raamsdonk JM, Gibson WT, Pearson J, Murphy Z, Lu G, Leavitt BR, et al. Body weight is modulated by levels of full-length huntingtin. Hum Mol Genet. 2006 May;15(9):1513–23.
129. Spampanato J, Gu X, Yang XW, Mody I. Progressive synaptic pathology of motor cortical neurons in a BAC transgenic mouse model of Huntington's disease. Neuroscience. 2008 Dec;157(3):606–20.
130. Abada YSK, Nguyen HP, Schreiber R, Ellenbroek B. Assessment of motor function, sensory motor gating and recognition memory in a novel BACHD transgenic rat model for huntington disease. PLoS One. 2013;8(7):e68584.
131. Southwell AL, Ko J, Patterson PH. Intrabody gene therapy ameliorates motor, cognitive, and neuropathological symptoms in multiple mouse models of Huntington's disease. J Neurosci. 2009 Oct;29(43): 13589–602.
132. Chen X, Wu J, Lvovskaya S, Herndon E, Supnet C, Bezprozvanny I. Dantrolene is neuroprotective in Huntington's disease transgenic mouse model. Mol Neurodegener. 2011 Nov;6:81.
133. Mantovani S, Gordon R, Li R, Christie DC, Kumar V, Woodruff TM. Motor deficits associated with Huntington's disease occur in the absence of striatal degeneration in BACHD transgenic mice. Hum Mol Genet. 2016 May;25(9):1780–91.
134. Hult S, Soylu R, Björklund T, Belgardt BF, Mauer J, Brüning JC, et al. Mutant huntingtin causes metabolic imbalance by disruption of hypothalamic neurocircuits. Cell Metab. 2011 Apr;13(4):428–39.
135. Vlamings R, Zeef DH, Janssen MLF, Oosterloo M, Schaper FD, Jahanshahi A, et al. Lessons learned from the transgenic Huntington's disease rats. Neural Plast. 2012;2012:682712.
136. von Hörsten S, Schmitt I, Nguyen HP, Holzmann C, Schmidt T, Walther T, et al. Transgenic rat model of Huntington's disease. Hum Mol Genet. 2003 Mar;12(6):617–24.
137. Nguyen HP, Kobbe P, Rahne H, Wörpel T, Jäger B, Stephan M, et al. Behavioral abnormalities precede neuropathological markers in rats transgenic for Huntington's disease. Hum Mol Genet. 2006 Nov;15(21):3177–94.
138. Carreira JC, Jahanshahi A, Zeef D, Kocabicak E, Vlamings R, von Hörsten S, et al. Transgenic rat models of Huntington's disease. Curr Top Behav Neurosci. 2015;22:135–47.
139. Kántor O, Temel Y, Holzmann C, Raber K, Nguyen HP, Cao C, et al. Selective striatal neuron loss and alterations in behavior correlate with impaired striatal function in Huntington's disease transgenic rats. Neurobiol Dis. 2006 Jun;22(3):538–47.
140. Zeef DH, Vlamings R, Lim LW, Tan S, Janssen MLF, Jahanshahi A, et al. Motor and non-motor behaviour in experimental Huntington's disease. Behav Brain Res. 2012 Jan;226(2):435–9.
141. Cao C, Temel Y, Blokland A, Ozen H, Steinbusch HWM, Vlamings R, et al. Progressive deterioration of reaction time performance and choreiform symptoms in a new Huntington's disease transgenic ratmodel. Behav Brain Res. 2006 Jun;170(2):257–61.
142. Bode FJ, Stephan M, Suhling H, Pabst R, Straub RH, Raber KA, et al. Sex differences in a transgenic rat model of Huntington's disease: Decreased 17beta-estradiol levels correlate with reduced numbers of DARPP32+ neurons in males. Hum Mol Genet. 2008 Sep;17(17):2595–609.
143. Kurat S, Heinrich P, Molnar-Kasza A, Loeffler T, Flunkert S, Hutter-Paier B. Homozygosity of BACHD rats not only causes strong behavioral deficits in young female rats but also a reduced breeding success. Brain Res. 2021 Mar;1761:147396.
144. Kirch RD, Meyer PT, Geisler S, Braun F, Gehrig S, Langen KJ, et al. Early deficits in declarative and procedural memory dependent behavioral function in a transgenic rat model of Huntington's disease. Behav Brain Res. 2013 Feb;239:15–26.
145. Ortiz AN, Osterhaus GL, Lauderdale K, Mahoney L, Fowler SC, von Hörsten S, et al. Motor function and dopamine release measurements in transgenic Huntington's disease model rats. Brain Res. 2012 Apr;1450:148–56.

146. Zeef DH, van Goethem NP, Vlamings R, Schaper F, Jahanshahi A, Hescham S, et al. Memory deficits in the transgenic rat model of Huntington's disease. Behav Brain Res. 2012 Feb;227(1):194–8.
147. Casteels C, Vandeputte C, Rangarajan JR, Dresselaers T, Riess O, Bormans G, et al. Metabolic and type 1 cannabinoid receptor imaging of a transgenic rat model in the early phase of Huntington disease. Exp Neurol. 2011 Jun;229(2):440–9.
148. Petrasch-Parwez E, Nguyen HP, Löbbecke-Schumacher M, Habbes HW, Wieczorek S, Riess O, et al. Cellular and subcellular localization of huntingtin [corrected] aggregates in the brain of a rat transgenic for Huntington disease. J Comp Neurol. 2007 Apr;501(5):716–30.
149. Winkler C, Gil JMAC, Araújo IM, Riess O, Skripuletz T, von Hörsten S, et al. Normal sensitivity to excitotoxicity in a transgenic Huntington's disease rat. Brain Res Bull. 2006 Apr;69(3):306–10.
150. Fink KD, Rossignol J, Crane AT, Davis KK, Bavar AM, Dekorver NW, et al. Early cognitive dysfunction in the HD 51 CAG transgenic rat model of Huntington's disease. Behav Neurosci. 2012 Jun;126(3):479–87.
151. Yu-Taeger L, Petrasch-Parwez E, Osmand AP, Redensek A, Metzger S, Clemens LE, et al. A novel BACHD transgenic rat exhibits characteristic neuropathological features of Huntington disease. J Neurosci. 2012 Oct;32(44):15426–38.
152. Clemens LE, Weber JJ, Wlodkowski TT, Yu-Taeger L, Michaud M, Calaminus C, et al. Olesoxime suppresses calpain activation and mutant huntingtin fragmentation in the BACHD rat. Brain. 2015 Dec;138(Pt 12):3632–53.
153. Yu-Taeger L, Bonin M, Stricker-Shaver J, Riess O, Nguyen HHP. Dysregulation of gene expression in the striatum of BACHD rats expressing full-length mutant huntingtin and associated abnormalities on molecular and protein levels. Neuropharmacology. 2017 May;117:260–72.
154. Jansson EKH, Clemens LE, Riess O, Nguyen HP. Reduced motivation in the BACHD rat model of Huntington disease is dependent on the choice of food deprivation strategy. PLoS One. 2014;9(8):e105662.
155. Menalled LB, Kudwa AE, Miller S, Fitzpatrick J, Watson-Johnson J, Keating N, et al. Comprehensive behavioral and molecular characterization of a new knock-in mouse model of Huntington's disease: zQ175. PLoS One. 2012;7(12):e49838.
156. Menalled LB. Knock-in mouse models of Huntington's disease. NeuroRx. 2005 Jul;2(3):465–70.
157. Wheeler VC, White JK, Gutekunst CA, Vrbanac V, Weaver M, Li XJ, et al. Long glutamine tracts cause nuclear localization of a novel form of huntingtin in medium spiny striatal neurons in HdhQ92 and HdhQ111 knock-in mice. Hum Mol Genet. 2000 Mar;9(4):503–13.
158. Woodman B, Butler R, Landles C, Lupton MK, Tse J, Hockly E, et al. The Hdh(Q150/Q150) knock-in mouse model of HD and the R6/2 exon 1 model develop comparable and widespread molecular phenotypes. Brain Res Bull. 2007 Apr;72(2–3):83–97.
159. Shelbourne PF, Killeen N, Hevner RF, Johnston HM, Tecott L, Lewandoski M, et al. A Huntington's disease CAG expansion at the murine Hdh locus is unstable and associated with behavioural abnormalities in mice. Hum Mol Genet. 1999 May;8(5):763–74.
160. Giralt A, Saavedra A, Alberch J, Pérez-Navarro E. Cognitive dysfunction in Huntington's disease: Humans, mouse models and molecular mechanisms. J Huntingtons Dis. 2012;1(2):155–73.
161. Orvoen S, Pla P, Gardier AM, Saudou F, David DJ. Huntington's disease knock-in male mice show specific anxiety-like behaviour and altered neuronal maturation. Neurosci Lett. 2012 Jan;507(2):127–32.
162. Stefanko DP, Shah VD, Yamasaki WK, Petzinger GM, Jakowec MW. Treadmill exercise delays the onset of non-motor behaviors and striatal pathology in the CAG(140) knock-in mouse model of Huntington's disease. Neurobiol Dis. 2017 Sep;105:15–32.
163. Wheeler VC, Gutekunst CA, Vrbanac V, Lebel LA, Schilling G, Hersch S, et al. Early phenotypes that presage late-onset neurodegenerative disease allow testing of modifiers in Hdh CAG knock-in mice. Hum Mol Genet. 2002 Mar;11(6):633–40.
164. Menalled LB, Sison JD, Dragatsis I, Zeitlin S, Chesselet MF. Time course of early motor and neuropathological anomalies in a knock-in mouse model of Huntington's disease with 140 CAG repeats. J Comp Neurol. 2003 Oct;465(1):11–26.
165. Hickey MA, Kosmalska A, Enayati J, Cohen R, Zeitlin S, Levine MS, et al. Extensive early motor and non-motor behavioral deficits are followed by striatal neuronal loss in knock-in Huntington's disease mice. Neuroscience. 2008 Nov;157(1):280–95.
166. Lerner RP, Trejo Martinez LDCG, Zhu C, Chesselet MF, Hickey MA. Striatal atrophy and dendritic alterations in a knock-in mouse model of Huntington's disease. Brain Res Bull. 2012 Apr;87(6):571–8.
167. Cummings DM, Cepeda C, Levine MS. Alterations in striatal synaptic transmission are consistent across genetic mouse models of Huntington's disease. ASN Neuro. 2010 Jun;2(3):e00036.

168. Heng MY, Tallaksen-Greene SJ, Detloff PJ, Albin RL. Longitudinal evaluation of the Hdh(CAG)150 knock-in murine model of Huntington's disease. J Neurosci. 2007 Aug;27(34):8989–98.
169. Lin CH, Tallaksen-Greene S, Chien WM, Cearley JA, Jackson WS, Crouse AB, et al. Neurological abnormalities in a knock-in mouse model of Huntington's disease. Hum Mol Genet. 2001 Jan;10(2):137–44.
170. Tallaksen-Greene SJ, Crouse AB, Hunter JM, Detloff PJ, Albin RL. Neuronal intranuclear inclusions and neuropil aggregates in HdhCAG(150) knockin mice. Neuroscience. 2005;131(4):843–52.
171. Brooks S, Higgs G, Jones L, Dunnett SB. Longitudinal analysis of the behavioural phenotype in Hdh(CAG)150 Huntington's disease knock-in mice. Brain Res Bull. 2012 Jun;88(2–3):182–8.
172. Rattray I, Smith EJ, Crum WR, Walker TA, Gale R, Bates GP, et al. Correlations of behavioral deficits with brain pathology assessed through longitudinal MRI and histopathology in the HdhQ150/Q150 mouse model of Huntington's disease. PLoS One. 2017;12(1):e0168556.
173. Kuhn A, Goldstein DR, Hodges A, Strand AD, Sengstag T, Kooperberg C, et al. Mutant Huntingtin's effects on striatal gene expression in mice recapitulate changes observed in human Huntington's disease brain and do not differ with mutant huntingtin length or wild-type huntingtin dosage. Hum Mol Genet. 2007 Aug;16(15):1845–61.
174. Menalled LB, Sison JD, Wu Y, Olivieri M, Li XJ, Li H, et al. Early motor dysfunction and striosomal distribution of huntingtin microaggregates in Huntington's disease knock-in mice. J Neurosci. 2002 Sep;22(18):8266–76.
175. Heng MY, Duong DK, Albin RL, Tallaksen-Greene SJ, Hunter JM, Lesort MJ, et al. Early autophagic response in a novel knock-in model of Huntington disease. Hum Mol Genet. 2010 Oct;19(19):3702–20.
176. Smith GA, Rocha EM, McLean JR, Hayes MA, Izen SC, Isacson O, et al. Progressive axonal transport and synaptic protein changes correlate with behavioral and neuropathological abnormalities in the heterozygous Q175 KI mouse model of Huntington's disease. Hum Mol Genet. 2014 Sep;23(17):4510–27.
177. Heikkinen T, Lehtimäki K, Vartiainen N, Puoliväli J, Hendricks SJ, Glaser JR, et al. Characterization of neurophysiological and behavioral changes, MRI brain volumetry and 1H MRS in zQ175 knock-in mouse model of Huntington's disease. PLoS One. 2012;7(12):e50717.
178. Peng Q, Wu B, Jiang M, Jin J, Hou Z, Zheng J, et al. Characterization of behavioral, neuropathological, brain metabolic and key molecular changes in zQ175 knock-in mouse model of Huntington's disease. PLoS One. 2016;11(2):e0148839.
179. Stricker-Shaver J, Novati A, Yu-Taeger L, Nguyen HP. Genetic rodent models of Huntington disease. Adv Exp Med Biol. 2018;1049:29–57.

7 Rodent Models of Prion Disease

Saeid Bagheri-Mohammadi

This chapter reviews the development, characteristics, and applications of rodent models in prion research, with emphasis on their use in studying strain-specific properties, transmission barriers, and tissue tropism. It also addresses the advantages and limitations of each model, including expression pattern differences, genetic background effects, and relevance to human pathology. By integrating molecular insights with in vivo experimentation, rodent models continue to provide an essential bridge between fundamental prion biology and translational research aimed at improving diagnostics, bioassays, and potential therapies.

7.1 INTRODUCTION

Prion diseases, also referred to as transmissible spongiform encephalopathies (TSEs), are a group of fatal neurodegenerative disorders caused by abnormally folded proteins known as prions. These conditions may present in sporadic, genetic, or acquired forms. Sporadic forms include Jakob–Creutzfeldt disease (JCD), genetic forms encompass genetic Jakob–Creutzfeldt disease (gJCD), Gerstmann–Sträussler–Scheinker syndrome (GSS), and fatal familial insomnia (FFI), while acquired forms include kuru, variant Jakob–Creutzfeldt disease (vJCD), and iatrogenic Jakob–Creutzfeldt disease (iaJCD).[1,2]

Prion diseases occur when the normal prion protein (PrP^C, where "C" denotes the cellular form), which mainly has an α-helical structure, is transformed into an abnormal form known as the prion (PrP^{Sc}, with "Sc" referring to scrapie—a prion disease affecting sheep and goats). This abnormal protein is a proteinaceous infectious particle characterized by a predominantly β-pleated sheet structure. The conversion of PrP^D to PrP^{sc} is a post-translational conformational change that not only alters the protein's secondary structure but also confers resistance to proteolytic degradation, leading to its accumulation in the central nervous system (CNS). Progressive deposition of PrP^{sc} triggers synaptic dysfunction, neuronal loss, spongiform changes, and reactive gliosis, ultimately culminating in rapid neurological decline and death. In addition to their unique protein-only infectious nature, prion diseases are notable for their long incubation periods, strain-specific properties, and the ability to cross species barriers under certain conditions.

7.2 HISTORICAL BACKGROUND

Between 1921 and 1923, Alfons Jakob documented the first human cases of prion disease and believed they were similar to a case involving a young woman described by Hans Creutzfeldt in 1920.[3,4] For many years, this illness was commonly known as Jakob or Jakob-Creutzfeldt disease. However, Clarence J. Gibbs, a leading researcher, began using the term Creutzfeldt-Jakob disease because the acronym more closely matched his own initials. Subsequent reanalysis revealed that only two of Jakob's five cases, and none of Creutzfeldt's, were consistent with what is now known as prion disease. Therefore, the correct name for the disease should be Jakob or Jakob-Creutzfeldt disease.[2] For decades, prion diseases were thought to be caused by "slow viruses" because of their transmissibility and long incubation periods. This hypothesis was challenged when Tikvah Alper

DOI: 10.1201/9781032676524-8

and Ian Pattison proposed that the scrapie agent might lack nucleic acids, and subsequent studies showed that protein-denaturing treatments could abolish infectivity.[5,6]

In 1997, Stanley Prusiner was awarded the Nobel Prize in Physiology or Medicine for isolating the scrapie agent and demonstrating it to be a "proteinaceous infectious particle," or prion. His prion hypothesis—that these infectious agents are composed solely of misfolded protein—was initially controversial but is now supported by overwhelming evidence. Prusiner demonstrated that prions propagate by inducing normally folded proteins to adopt the abnormal, infectious conformation, thereby explaining both the transmission and replication mechanisms of prion diseases. This groundbreaking discovery not only introduced a new class of infectious agents but also transformed the scientific understanding of several neurodegenerative disorders.[7]

7.3 MOLECULAR PATHOGENESIS

Prion diseases are unique in medicine because they can develop in three different ways:

- Sporadically (without a known cause)
- Genetically (due to mutations in the prion protein (PRNP) gene)
- Acquired (through exposure to infectious prions)

The underlying mechanism involves the abnormal misfolded form of the prion protein, called PrP^{Sc} (with "Sc" referring to scrapie, a prion disease found in sheep and goats), which acts as a template. When PrP^{Sc} comes into contact with the normal cellular form of the prion protein, PrP^{C}, it converts PrP^{C} into the diseased PrP^{Sc} form, thereby producing two prions.[2] Two prions convert two more PrP^{C} molecules into PrP^{Sc}, which then convert four additional PrP^{C} molecules, continuing this exponential process and causing a buildup of prions. There is ongoing debate about whether neurodegeneration is caused by the accumulation of PrP^{Sc} itself or potentially by an intermediate form of the prion that appears during the conversion from PrP^{C} to PrP^{Sc}.[2,8] The PrPsc form is partially resistant to breakdown by proteases and serves as a template that induces the normal PrP^{C} proteins to also misfold into the abnormal PrP^{SC} shape.[9–11] Each PrP^{SC} molecule transforms a normal protein into its abnormal form, resulting in a rapid and exponential increase in the abnormal protein. This conversion, along with the buildup and clumping of PrP^{SC}, is linked to the death of neurons and the characteristic spongy appearance of the brain seen in prion diseases, which includes loss of nerve cells, gliosis, and the formation of vacuoles.[1]

7.4 CLINICAL SPECTRUM AND DIAGNOSIS

The most common human prion disease is sporadic CJD (sCJD), followed by genetic prion diseases such as gCJD, GSS, and FFI, and then acquired forms (kuru, iaCJD, vCJD).[2,12]

Human prion diseases can be difficult to diagnose due to their diverse clinical presentations. However, advances in testing methods like real-time quaking-induced conversion (RT-QuIC) and the use of immunohistochemistry or western blotting to detect PrPsc deposits have enhanced diagnostic precision. Additional tools include EEG and MRI, which can support the diagnosis of sCJD.

Given the broad differential diagnosis for rapidly progressive dementia, it is essential to remain alert to the possibility of prion disease despite its rarity.[2,12] These disorders are invariably fatal and currently untreatable, yet diagnosis is important for genetic counseling and for preventing acquired transmission. Fortunately, acquired forms such as kuru, iaCJD, and vCJD—though historically the most concerning—have been nearly eliminated through effective preventive measures. Nonetheless, vigilance remains necessary due to their long incubation periods and the potential emergence of new prion diseases from animal sources.

In contrast, the less common sporadic and genetic forms (including gCJD, GSS, FFI, and variably protease-sensitive prionopathy [VPSPr]) show variable diagnostic sensitivity, and RT-QuIC has not been validated for all of these conditions.[1] In suspected cases where other mimicking conditions have been excluded, diagnosis ultimately requires either genetic testing or neuropathological confirmation at autopsy.

7.5 GENETIC FORMS OF PRION DISEASE

Mutations in the prion protein gene (PRNP, located on chromosome 20p13) cause genetic prion diseases. Most pathogenic variants are missense mutations, though octapeptide repeat insertions (OPRI) and rare truncating/frameshift mutations also occur.[13,14] Many PRNP variants once thought to be disease-causing have since been shown to carry little or no risk. However, some variants are highly penetrant, as evidenced by their greater frequency in prion disease patients versus the general population, along with a higher percentage of cases having a family history of the disease.[15] The most frequently occurring highly penetrant variants include E200K, P102L, D178N, 6-OPRI, 5-OPRI, A117V, and P105L.[15] All of these variants fulfill the American College of Medical Genetics and Genomics/Association for Molecular Pathology (ACMG/AMP) criterion PS4, which indicates they are more commonly found in affected individuals compared to controls.[16] Twenty more variants have been identified as likely highly penetrant based on a combination of ACMG/AMP criteria: PM2 (absence from controls in a large population dataset) along with either PS2 (de novo variant reported in a patient without a family history) or PP1 (variant cosegregates with disease in a family with multiple affected members). Additionally, a few less penetrant variants, such as V210I, V180I, and M232R, seem to be associated with a lifetime risk ranging from 0.1% to 10%.[15]

Besides these disease-causing variants, the PRNP gene also contains a common sequence variation at codon 129, where either methionine (M) or valine (V) can be present. The genotype at codon 129 affects the length of the disease in certain genetic prion disorders and also impacts the risk of developing sporadic and iatrogenic prion diseases.[14] This codon does not seem to affect the age at which symptoms begin for the most common genotypes in genetic prion disease,[17] although it might have an impact in uncommon cases. The pathogenic variants mentioned earlier are linked to a range of prion disease manifestations. For example, variants like E200K, D178N-129V, and V210I have commonly been referred to as gCJD. On the other hand, P102L, P105L, and A117V were traditionally connected to GSS, while the D178N-129M variant was associated with FFI. These terms, CJD, GSS, and FFI, were established before the identification of PRNP as the single gene responsible for all prion diseases. Today, it is understood that there is considerable variability in clinical features, with overlaps among these historically defined categories.[18]

Since the prion protein is crucial for disease transmission and neurotoxicity, numerous experimental therapies aimed at reducing its production are being explored and show potential as disease-modifying treatments for all types of prion diseases, even in individuals who carry mutations but have no symptoms.[13]

7.6 NEUROINVASION AND MECHANISMS OF TOXICITY

Although most TSEs occur spontaneously, the infectious transmission of these diseases has attracted significant scientific and public attention. In laboratory settings, the TSE agent is typically introduced via direct injection into the brain of experimental animals; however, natural transmission most commonly occurs through ingestion. For example, strong evidence indicates that the bovine spongiform encephalopathy (BSE) outbreak in England was caused by feeding cattle meat and bone meal contaminated with BSE-infected material. Humans consuming infected cattle products are believed to have developed vCJD as a consequence.[19–21] Neuroinvasion usually begins following oral exposure to the TSE agent. Experimental data suggest that this process involves transcytosis through intestinal microfold (M) cells.[22,23] Migratory dendritic cells may also directly sample antigens from the intestinal lumen, potentially facilitating initial uptake of the prion agent. Once the abnormal prion protein (PrP^{Sc}) crosses the epithelial barrier, it is engulfed by antigen-presenting cells such as macrophages and dendritic cells. While macrophages appear to exert a protective effect, dendritic cells are implicated in transporting PrP^{Sc} to follicular dendritic cells within the germinal centers of Peyer's patches and other gut-associated lymphoid tissues (GALT). Following replication in these lymphoid tissues and the spleen, the agent spreads to the CNS via the enteric nervous system. This retrograde neuroinvasion occurs along both sympathetic (e.g., splanchnic nerve) and parasympathetic (e.g., vagus nerve) efferent fibers.[22,23]

The precise mechanism of prion transport between synaptically connected peripheral neurons remains unclear. Possible pathways include sequential interactions involving direct contact between PrP^{Sc} and normal prion protein (PrP^{C}) on axonal membranes, vesicle-mediated transfer, or dissemination via extracellular aggregates.[22] After entering the CNS, the agent continues its backward propagation, ultimately infecting the brain and causing characteristic spongiform degeneration alongside astroglial activation.

Several cellular factors have been implicated in facilitating neuroinvasion. Notably, ferritin associates with PrP^{Sc} and mediates its endocytosis and vesicular transcytosis in intestinal epithelial cells.[24] Components of the complement system, including C1q, C2, C3, and factor B, may promote initial transport of PrP^{Sc} to lymphoreticular cells. Polyanionic glycans such as heparan sulfate serve as cell surface receptors involved in prion uptake, influencing PrP conversion and propagation. Moreover, expression of the laminin receptor (or its precursor), a glycosaminoglycan-binding protein, is essential for prion infectivity in cell culture models and contributes to both uptake and spread of the agent.

Despite these insights, the identity of the toxic prion species and the exact mechanisms underlying neurotoxicity remain unresolved. TSE pathology does not seem to result primarily from loss of normal PrP^{C} function, as mice lacking the PrP gene exhibit only mild cognitive deficits.[25] Intriguingly, the degree of PrP^{Sc} accumulation does not always correlate with the extent of neurodegeneration. For instance, transgenic mice expressing mutant PrP spontaneously develop neurological disease without producing infectious aggregates, whereas mice expressing anchorless PrP accumulate highly infectious aggregates but suffer less neurodegeneration than normal infected mice. These findings suggest that infectious and neurotoxic PrP species may be distinct.[25] The harmful form of PrP might represent an intermediate or byproduct in the conversion process from PrP^{C} to PrP^{Sc}. Similar to other neurodegenerative disorders, small oligomeric assemblies are suspected intermediates that may disrupt cell membranes by forming pore-like ring structures.[26] Although further research is needed to elucidate the precise structure of the neurotoxic species, recent studies propose several potential mechanisms of neuronal death in TSEs, including enhanced NMDA receptor-mediated excitotoxicity and activation of the Erk1/2 signaling pathway.[27,28]

7.7 EXPERIMENTAL MODELS TO STUDY PRION DISEASE PATHOGENESIS

Prion diseases result from the conversion of the normal host protein, PrP^{C}, into a pathogenic form known as PrP^{Sc}. Advances in biology, genetics, and prion research have enabled the study of this disease process using recombinant proteins, ex vivo systems, and in vitro models. However, mammalian hosts remain the gold standard for assessing prion infectivity, transmission, and strain variation. Despite lacking nucleic acids, prions encode strain-specific information through the unique conformation of their infectious protein, which influences disease progression, host susceptibility, and therapeutic outcomes.[29] To thoroughly investigate the strain characteristics of natural prions and develop effective treatments, it is essential to employ experimental mammalian models that faithfully replicate these infectious agents.

7.7.1 Rodent Models: Hamsters, Mice, and Bank Voles

Since the 1970s, Syrian golden hamsters have been widely used in prion research, particularly for studying the anatomical and functional routes of prion spread both peripherally and within the CNS.[30,31] Hamsters helped reveal unique strain characteristics and species barriers.[32] However, since the 1990s, genetically modified mouse models have largely supplanted hamsters, thanks to advances in transgenesis, gene knock-out, and inducible Prnp gene manipulation techniques. Although hamsters are occasionally still used for peripheral exposure studies,[33] the versatility and availability of mouse models have made hamster models largely obsolete.

More recently, bank voles have emerged as valuable prion models.[34,35] Their susceptibility to prions from various species is influenced by the amino acid at codon 109 of their PrP, which can be methionine or isoleucine. Mouse models expressing bank vole PrP with isoleucine at codon 109 spontaneously develop neurodegenerative disease, likely due to a natural propensity for misfolding.[36]

Bank voles efficiently transmit sporadic CJD prions (notably MM1, MV1, and MM2 subtypes, excluding MV2 and VV2 per Parchi classification),[37] as well as inherited prion diseases (such as V210I, E220K, A117V, P102L, and F198S), with transmission efficiency comparable to humanized transgenic mice.[35,38] Importantly, bank voles maintain consistent neuropathological and biochemical strain features.[35] Nonetheless, the modeled inherited prion diseases in bank voles show less resemblance to the neuropathology and biochemistry of human cases.[39]

7.7.1.1 Mouse Models: Milestones in Prion Research

Mouse models first contributed to prion research through serial transmission and adaptation of sheep scrapie in wild-type mice, allowing studies on species barriers and prion strain properties. Early experiments used inbred mice, but the field advanced dramatically with the advent of transgenic mouse technology.

The first transgenic mice expressing hamster PrP, developed in 1989[40] (several years after the first transgenic animals in 1982[41]), represented a breakthrough by overcoming species barriers and demonstrating the key role of Prnp gene sequence in prion susceptibility.

Early efforts to model inherited prion diseases, such as GSS syndrome, involved expressing a mouse analog of the human P102L mutation.

An important early effort to model inherited prion disease associated with the GSS syndrome involved expressing a mouse version of the human P102L mutation. These mice exhibited neurodegeneration but did not show the distinctive prion protein deposits seen in human GSS cases. Brain extracts from GSS mice transmitted disease within the same strain but failed to infect standard wild-type CD1 mice.[42,43] Co-expression of human transgenic Prnp alongside endogenous mouse Prnp led to interference effects between the prion proteins.

A pivotal development came in 1992 with the creation of Prnp knock-out mice.[44] These mice proved the essential role of normal host PrP in prion propagation[45] and provided a platform to express various transgenes without interference from endogenous PrP. Weissmann, who developed this model, famously called it "a mouse to remember." The cost-effectiveness of maintaining mice and advances in gene editing have cemented their status as the preferred model in prion research and broader life sciences. New mouse models continue to be generated to study prion strain biology, PrP function, transmission barriers, and to replicate human prion diseases. Subsequent sections will explore wild-type, knock-out, transgenic, and gene-targeted mice in modeling human prion diseases and PrP's physiological roles.[46]

7.7.1.2 Transgenic Mice as Ultra-Sensitive Biological Prion Detection Models

According to the prion hypothesis, prion infectivity should be detectable in transgenic mice expressing the PrP^C protein from the donor species with similar efficiency as in the original host. The highest transmission rates have been observed in transgenic mice that overexpress PrP^C, often due to multiple randomly inserted copies of the transgene introduced via pronuclear microinjection. However, this random insertion can cause PrP^C expression patterns that do not faithfully replicate the natural distribution seen in conventional mice, potentially affecting disease development.

To minimize the effects of random integration, researchers use large transgene constructs such as the half-genomic (hg-PrP) vector, a 12-kb mouse prion gene lacking the second intron, or a

45-kb cosmid containing the entire hamster prion gene. Alternatively, promoters from cytomegalovirus (CMV)[47,48] or neuron-specific enolase (NSE) genes[49] are employed to drive PrP^C expression. Nevertheless, these transgenic mice generally do not exhibit tissue expression patterns closely resembling those of wild-type mice.

The susceptibility of any transgenic mouse line is best evaluated by performing side-by-side titrations of a known inoculum in both the transgenic mice and the donor species. Notably, PrP^C overexpression in transgenic mice can confer greater sensitivity than the natural host. Comparative bioassays for CJD in monkeys[50] and BSE in cattle[51,52] have shown that transgenic mouse assays detect infectivity levels approximately 10–100 times higher (1–2 log LD_50) than in the original species.

Among ruminant prion models, Tgbov XV mice uniquely have such parallel titration data. Intracerebral inoculation of these mice with serial dilutions of a BSE brainstem homogenate yielded an infectivity titer of $10^{7.8}$ ID_{50}/g, whereas cattle inoculated with the same homogenate had a titer of 10^6 ID_{50}/g.

Because titrations in the original hosts are not always feasible, titration in a well-characterized transgenic mouse line can serve as a reference to compare other transgenic models. Unfortunately, such standardized data are lacking for most transgenic lines. Therefore, establishing parallel titration standards between original hosts and transgenic mice remains a key objective to better interpret infectivity data across tissues and fluids from prion-infected animals and humans.[53] Assessing transgenic mouse susceptibility is especially important given recent contradictory findings challenging the amino acid homology hypothesis. For example, a comparative study of sheep-passaged BSE infectivity in Romney sheep and RIII mice found the species barrier from sheep to mice to be lower than expected, with similar infectivity levels measured in sheep [$10^{5.4}$ (intracerebral) ID_{50}/g] and mice [$10^{5.0}$ (intracerebral + intraperitoneal) ID_{50}/g].[54]

It has been proposed that the enhanced susceptibility of transgenic mice may result more from overexpression of heterologous PrP^C than from sequence homology effects. However, tga20 mice, which overexpress murine PrP^C, showed no significant reduction in incubation period when challenged with BSE.[51,55] Addressing these questions is crucial not only for identifying the most sensitive models for detecting human and animal prions but also for gaining deeper insight into the mechanisms of prion replication.

7.8 CONCLUSION

Prion diseases remain among the most perplexing and lethal disorders in neurology, uniting the concepts of infectious disease, protein misfolding, and neurodegeneration in a single pathological process. Over the past century, advances in molecular biology, genetics, and experimental modeling have transformed our understanding of their origin, transmission, and cellular effects. Yet, despite remarkable progress, these diseases remain uniformly fatal, and the precise mechanisms of neurotoxicity are still unresolved.

The refinement of diagnostic techniques and the development of sensitive animal models have not only deepened our insight into disease mechanisms but also provided essential tools for assessing potential therapies. The emerging strategies aimed at lowering prion protein expression or interfering with its pathological conversion hold promise, particularly for presymptomatic individuals carrying high-risk PRNP mutations.

Continued vigilance is essential, both to prevent acquired forms from re-emerging and to detect novel animal-to-human transmissions. Equally important is sustained investment in basic and translational research, as lessons learned from prion biology extend beyond these rare disorders, offering critical clues to more common neurodegenerative diseases that share protein misfolding as a core mechanism. By bridging molecular insights with therapeutic innovation, the goal of altering the course of these invariably fatal diseases may one day be within reach.

REFFRENCES

1. Baldwin KJ, Correll CM. Prion disease. Seminars in Neurology [Internet]. 2019 [cited 2025 Aug 2];39(4):428–39. Available from: https://pubmed.ncbi.nlm.nih.gov/31533183/
2. Geschwind MD. Prion diseases. Continuum Lifelong Learning in Neurology [Internet]. 2015 Dec 1 [cited 2025 Aug 2];21(6):1612–38. Available from: https://pubmed.ncbi.nlm.nih.gov/26633779/
3. Creutzfeldt HG. On a particular focal disease of the central nervous system (preliminary communication). Alzheimer Disease & Associated Disorders. 1989;3(1):15–25.
4. Jakob A. Concerning a disorder of the central nervous system clinically resembling multiple sclerosis with remarkable anatomic findings (spastic pseudosclerosis): report of a fourth case. Alzheimer Disease & Associated Disorders. 1989;3(1):37–45.
5. Bechtel K, Geschwind MD. Ethics in prion disease. Progress in Neurobiology. 2013;110:29–44.
6. Prusiner SB. Prions. Proceedings of the National Academy of Sciences. 1998;95(23):13363–83.
7. Soto C. Prion hypothesis: the end of the controversy? Trends in Biochemical Sciences. 2011;36(3):151–8.
8. Moreno JA, Halliday M, Molloy C, Radford H, Verity N, Axten JM, et al. Oral treatment targeting the unfolded protein response prevents neurodegeneration and clinical disease in prion-infected mice. Science Translational Medicine. 2013;5(206):206ra138.
9. Pan KM, Baldwin M, Nguyen J, Gasset M, Serban ANA, Groth D, et al. Conversion of alpha-helices into beta-sheets features in the formation of the scrapie prion proteins. Proceedings of the National Academy of Sciences. 1993;90(23):10962–6.
10. Prusiner SB. Novel proteinaceous infectious particles cause scrapie. Science. 1982;216(4542):136–44.
11. Prusiner SB. Neurodegenerative diseases and prions. New England Journal of Medicine. 2001;344(20):1516–26.
12. Ironside JW, Ritchie DL, Head MW. Prion diseases. Handbook of Clinical Neurology. 2018;145:393–403.
13. Appleby BS, Shetty S, Elkasaby M. Genetic aspects of human prion diseases. Frontiers in Neurology. 2022;13:1003056.
14. Mead S, Lloyd S, Collinge J. Genetic factors in mammalian prion diseases. Annual Review of Genetics. 2019;53(1):117–47.
15. Minikel EV, Vallabh SM, Lek M, Estrada K, Samocha KE, Sathirapongsasuti JF, et al. Quantifying prion disease penetrance using large population control cohorts. Science Translational Medicine. 2016;8(322):322ra9.
16. Richards S, Aziz N, Bale S, Bick D, Das S, Gastier-Foster J, et al. Standards and guidelines for the interpretation of sequence variants: a joint consensus recommendation of the American College of Medical Genetics and Genomics and the Association for Molecular Pathology. Genetics in Medicine. 2015;17(5):405–23.
17. Minikel EV, Vallabh SM, Orseth MC, Brandel JP, Haïk S, Laplanche JL, et al. Age at onset in genetic prion disease and the design of preventive clinical trials. Neurology. 2019;93(2):e125–34.
18. Goldman JS, Vallabh SM. Genetic counseling for prion disease: updates and best practices. Genetics in Medicine. 2022;24(10):1993–2003.
19. Aguzzi A, Polymenidou M. Mammalian prion biology: one century of evolving concepts. Cell. 2004;116(2):313–27.
20. Collinge J. Prion diseases of humans and animals: their causes and molecular basis. Annual Review of Neuroscience. 2001;24(1):519–50.
21. Wells GAH, Wilesmith JW. The neuropathology and epidemiology of bovine spongiform encephalopathy. Brain Pathology. 1995;5(1):91–103.
22. Heikenwalder M, Julius C, Aguzzi A. Prions and peripheral nerves: a deadly rendezvous. Journal of Neuroscience Research. 2007;85(12):2714–25.
23. Mabbott NA, MacPherson GG. Prions and their lethal journey to the brain. Nature Reviews Microbiology. 2006;4(3):201–11.
24. Mishra RS, Basu S, Gu Y, Luo X, Zou WQ, Mishra R, et al. Protease-resistant human prion protein and ferritin are cotransported across Caco-2 epithelial cells: implications for species barrier in prion uptake from the intestine. Journal of Neuroscience. 2004;24(50):11280–90.
25. Cobb NJ, Surewicz WK. Prion diseases and their biochemical mechanisms. Biochemistry. 2009;48(12):2574–85.
26. Caughey B, Lansbury Jr PT. Protofibrils, pores, fibrils, and neurodegeneration: separating the responsible protein aggregates from the innocent bystanders. Annual Review of Neuroscience. 2003;26(1):267–98.

27. LaCasse RA, Striebel JF, Favara C, Kercher L, Chesebro B. Role of Erk1/2 activation in prion disease pathogenesis: absence of CCR1 leads to increased Erk1/2 activation and accelerated disease progression. Journal of Neuroimmunology. 2008;196(1–2):16–26.
28. Ratté S, Prescott SA, Collinge J, Jefferys JGR. Hippocampal bursts caused by changes in NMDA receptor-dependent excitation in a mouse model of variant CJD. Neurobiology of Disease. 2008;32(1):96–104.
29. DeFranco JP, Telling GC. The evolution of experimental rodent models for prion diseases. Journal of Neurochemistry. 2025;169(3):e70039.
30. Jendroska K, Heinzel FP, Torchia M, Stowring L, Kretzschmar HA, Kon A, et al. Proteinase-resistant prion protein accumulation in Syrian hamster brain correlates with regional pathology and scrapie infectivity. Neurology. 1991;41(9):1482.
31. Kimberlin RH, Walker CA. The role of the spleen in the neuroinvasion of scrapie in mice. Virus Research. 1989;12(3):201–11.
32. Kimberlin RH, Cole S, Walker CA. Temporary and permanent modifications to a single strain of mouse scrapie on transmission to rats and hamsters. Journal of General Virology. 1987;68(7):1875–81.
33. Chen B, Soto C, Morales R. Peripherally administrated prions reach the brain at sub-infectious quantities in experimental hamsters. FEBS Letters. 2014;588(5):795–800.
34. Cartoni C, Schinina ME, Maras B, Nonno R, Vaccari G, Di Bari MA, et al. Identification of the pathological prion protein allotypes in scrapie-infected heterozygous bank voles (*Clethrionomys glareolus*) by high-performance liquid chromatography–mass spectrometry. Journal of Chromatography A. 2005; 1081(1):122–6.
35. Nonno R, Bari MA Di, Cardone F, Vaccari G, Fazzi P, Dell'Omo G, et al. Efficient transmission and characterization of Creutzfeldt–Jakob disease strains in bank voles. PLoS Pathogens. 2006;2(2):e12.
36. Watts JC, Giles K, Stöhr J, Oehler A, Bhardwaj S, Grillo SK, et al. Spontaneous generation of rapidly transmissible prions in transgenic mice expressing wild-type bank vole prion protein. Proceedings of the National Academy of Sciences. 2012;109(9):3498–503.
37. Parchi P, Giese A, Capellari S, Brown P, Schulz-Schaeffer W, Windl O, et al. Classification of sporadic Creutzfeldt-Jakob disease based on molecular and phenotypic analysis of 300 subjects. Annals of Neurology. 1999;46(2):224–33.
38. Pirisinu L, Di Bari MA, D'Agostino C, Marcon S, Riccardi G, Poleggi A, et al. Gerstmann-Sträussler-Scheinker disease subtypes efficiently transmit in bank voles as genuine prion diseases. Scientific Reports. 2016;6(1):20443.
39. Watts JC, Giles K, Bourkas MEC, Patel S, Oehler A, Gavidia M, et al. Towards authentic transgenic mouse models of heritable PrP prion diseases. Acta Neuropathologica. 2016;132(4):593–610.
40. Scott M, Foster D, Mirenda C, Serban D, Coufal F, Wälchli M, et al. Transgenic mice expressing hamster prion protein produce species-specific scrapie infectivity and amyloid plaques. Cell. 1989;59(5): 847–57.
41. Palmiter RD, Brinster RL, Hammer RE, Trumbauer ME, Rosenfeld MG, Birnberg NC, et al. Dramatic growth of mice that develop from eggs microinjected with metallothionein–growth hormone fusion genes. Nature. 1982;300(5893):611–5.
42. Hsiao KK, Groth D, Scott M, Yang SL, Serban H, Rapp D, et al. Serial transmission in rodents of neurodegeneration from transgenic mice expressing mutant prion protein. Proceedings of the National Academy of Sciences. 1994;91(19):9126–30.
43. Hsiao KK, Scott M, Foster D, Groth DF, DeArmond SJ, Prusiner SB. Spontaneous neurodegeneration in transgenic mice with mutant prion protein. Science. 1990;250(4987):1587–90.
44. Büeler H, Fischer M, Lang Y, Bluethmann H, Lipp HP, DeArmond SJ, et al. Normal development and behaviour of mice lacking the neuronal cell-surface PrP protein. Nature. 1992;356(6370):577–82.
45. Sailer A. No propagation of prions in mice devoid of PrP. Cell. 1994;77(7):967–8.
46. Brandner S, Jaunmuktane Z. Prion disease: experimental models and reality. Acta Neuropathologica. 2017;133(2):197–222.
47. Scott MR, Köhler R, Foster D, Prusiner SB. Chimeric prion protein expression in cultured cells and transgenic mice. Protein Science. 1992;1(8):986–97.
48. Vilotte JL, Laude H. Transgenesis applied to transmissible spongiform encephalopathies. Transgenic Research. 2002;11:547–64.
49. Crozet C, Flamant F, Bencsik A, Aubert D, Samarut J, Baron T. Efficient transmission of two different sheep scrapie isolates in transgenic mice expressing the ovine PrP gene. Journal of Virology. 2001;75(11):5328–34.

50. Safar JG, Geschwind MD, Deering C, Didorenko S, Sattavat M, Sanchez H, et al. Diagnosis of human prion disease. Proceedings of the National Academy of Sciences. 2005;102(9):3501–6.
51. Buschmann A, Groschup MH. Highly bovine spongiform encephalopathy–sensitive transgenic mice confirm the essential restriction of infectivity to the nervous system in clinically diseased cattle. The Journal of Infectious Diseases. 2005;192(5):934–42.
52. Wells GAH, Konold T, Arnold ME, Austin AR, Hawkins SAC, Stack M, et al. Bovine spongiform encephalopathy: the effect of oral exposure dose on attack rate and incubation period in cattle. Journal of General Virology. 2007;88(4):1363–73.
53. Groschup M, Buschmann A. Rodent models for prion diseases. Veterinary Research. 2008;39(4):1–13.
54. Gonzalez L, Chianini F, Martin S, Sisó S, Gibbard L, Reid HW, et al. Comparative titration of experimental ovine BSE infectivity in sheep and mice. Journal of General Virology. 2007;88(2):714–7.
55. Buschmann A, Pfaff E, Reifenberg K, Müller HM, Groschup MH. Detection of cattle-derived BSE prions using transgenic mice overexpressing bovine PrP C. Springer; 2000.

8 Behavioral Assays in Rodent Models

Methods for Assessing Motor, Cognitive, and Social Phenotypes

Fatemeh Abbaszadeh, Fariba Akhondzadeh, and Pegah Javadpour

8.1 INTRODUCTION

Behavioral assays in rodent models are fundamental tools in neuroscience research, offering vital insights into the neurobiological mechanisms underlying cognition, emotion, social interaction, and their dysregulation in neurological and psychiatric disorders. Rodents, particularly mice and rats, are extensively employed due to their genetic tractability, well-characterized neurobiology, and capacity to exhibit complex, human-like behaviors (1,2). Their relatively short lifespans, ease of handling, and compatibility with advanced neuroimaging, electrophysiological, and molecular techniques make them indispensable for translational research bridging basic neuroscience and clinical applications.

This chapter provides a comprehensive overview of widely used behavioral paradigms in rodents, categorized by functional domains. It begins with cognitive tests such as the Morris water maze (MWM), Barnes maze, T-maze, Y-maze, object location test (OLT), and NOR test, which are designed to evaluate learning, memory, and decision-making capacities in rodents (3). These paradigms are sensitive to hippocampal and cortical function and are particularly valuable for investigating neurodevelopmental and neurodegenerative models.

Social behavior assessments, such as the reciprocal social interaction test and the three-chamber social interaction test, are presented as tools for investigating social motivation, recognition, and affiliative behaviors in both normative and disease states (4).

The chapter also reviews models of depressive- and amnesia-like behavior, including the forced swim test, tail suspension test, and sucrose preference test, which are widely used to evaluate behavioral despair and anhedonia in models of affective disorders. While often criticized for their ethological validity, they remain among the most robust and reproducible measures for screening antidepressant efficacy in rodents. Finally, the chapter explores motor function evaluations, including the rotarod test, beam walk test, and pole test, which are essential for assessing coordination, balance, and locomotor performance, particularly in models of neurodegenerative diseases such as Parkinson's disease, Huntington's disease, and amyotrophic lateral sclerosis. They can sensitively detect subtle motor impairments, track disease progression, and measure functional recovery following experimental treatments.

Collectively, these behavioral assays provide an integrative framework for characterizing phenotypic outcomes associated with genetic modifications, environmental manipulations, and pharmacological interventions. They not only enable the modeling of specific dimensions of human neuropsychiatric conditions but also facilitate the development and evaluation of targeted therapeutic

DOI: 10.1201/9781032676524-9

strategies. This chapter aims to provide a detailed methodological guide and conceptual overview of these tests, underscoring their critical role in preclinical behavioral neuroscience research.

This chapter aims to provide a detailed methodological guide and conceptual overview of these tests, underscoring their critical role in preclinical behavioral neuroscience research. In doing so, it highlights the technical considerations and interpretative nuances associated with each assay.

8.2 COGNITIVE FUNCTION TESTS

8.2.1 Maze Tests

Maze tests are commonly employed in behavioral neuroscience to evaluate cognitive functions like spatial learning, memory, decision-making, and anxiety in laboratory animals, particularly rodents. These assessments typically involve placing animals in structured environments, such as the Y-maze, T-maze, Barnes maze, or MWM, where they must navigate to locate food, find an exit, or reach a safe zone using spatial cues. The design of each maze (including arms, platforms, or circular pools) specifies the particular cognitive or emotional behaviors being investigated. Researchers analyze data such as time spent, paths taken, or mistakes made to assess memory, learning, decision-making, and anxiety-like behaviors. Maze tests offer valuable insights into the neural mechanisms that underpin cognition and have significant applications in studies of neurodegenerative diseases, drug effects, and brain injuries (Table 8.1).

8.2.1.1 MWM Test

Among the various types of water mazes used in behavioral research, the most widely recognized is the MWM, developed by Richard Morris (5). This paradigm was specifically designed to evaluate spatial learning by assessing an animal's ability to learn and recall the location of a submerged escape platform within a circular pool.

The MWM is a highly effective tool for detecting a range of cognitive changes, including those induced by drugs, brain lesions, or genetic manipulations that impact learning and memory processes (6). Also, its reliability and effectiveness have been consistently demonstrated across various preclinical models, including mice, rats, and guinea pigs (6).

TABLE 8.1
Comparison of Common Maze Tests in Behavioral Neuroscience

Maze Type	Structure	Purpose	Stress Level	Typical Measured Outcomes
Y-maze	Three identical arms at 120° angles forming a "Y"	Spontaneous alternation; spatial working memory	Low stress; controlled environment, relies on natural exploration and curiosity	Number of arm entries, percentage of alternation, time spent per arm, preference for novel arm
T-maze	Two arms at 180° angles, forming a "T"	Spatial memory; decision-making	Low stress; controlled environment, relies on natural exploration and curiosity	Correct arm choices, latency to choose, error rate
Barnes maze	Elevated circular platform with ~20 holes; one leads to an escape box	Spatial learning and memory	Moderate stress; uses aversion to open spaces	Number of errors, latency to find the escape box, and path length
Morris water maze	Large circular pool filled with opaque water; hidden platform	Spatial learning and memory; hippocampal function	High stress; uses aversion to water and swimming	Escape latency, path length, swimming speed, and time spent in target quadrant

Various protocol adaptations of the MWM allow for the assessment of different aspects of cognition, such as spatial navigation, working memory, latent learning, discrimination learning, and cued learning. These variations make the MWM a versatile and widely applicable paradigm in neuroscience research (Vorhees & Williams, 2006). The effectiveness of the MWM is closely linked to fundamental neurobiological mechanisms of memory, particularly long-term potentiation (LTP) and N-methyl-D-aspartate (NMDA) receptor function within the hippocampus. While the hippocampus plays a central role in spatial learning assessed by the MWM, several other brain regions also contribute to task performance. These include the entorhinal, perirhinal, prefrontal, and cingulate cortices, as well as the neostriatum. Additionally, the cerebellum may be involved to a limited extent, particularly in aspects related to procedural components of the task (7).

During the task, the animal is required to locate a small, hidden platform that is submerged just below the water's surface and remains in a fixed position throughout the trials. This platform serves as the goal, and the animal must use spatial memory to navigate toward it by relying on distal environmental cues.

As it was mentioned, the primary application of the MWM is to assess spatial, or place learning. In this paradigm, proximal cues such as nearby objects, textures, or odors are minimized or eliminated to prevent the animal from using them for navigation. Instead, the task is designed so that the animal must rely on distal cues, such as visual markers placed on the walls of the testing room, to orient itself and locate the hidden platform. This configuration ensures that the task specifically measures true spatial memory.

- **Protocol of the Test**
  - Training Protocols in the MWM
    - Daily Training Sessions and Trial Repetition

      Animals are generally trained across multiple trials per day to support the gradual acquisition of spatial memory. A widely adopted approach consists of four trials per day, allowing subjects repeated opportunities to learn the location of the hidden platform.
    - Start Point Variation

      To prevent the development of motor-based or procedural strategies and to ensure reliance on spatial cues, each training trial begins from a different location around the perimeter of the pool. These start positions are selected using a semi-randomized sequence; all four designated points are used once per session, but the order is varied across days and subjects to reduce predictability.
    - Inter-Trial Intervals (ITIs)

      The interval between consecutive trials within a single training session can vary based on experimental aims:
      - *Short ITIs*: Typically range from 10 to 15 seconds, allowing rapid succession of trials and minimal delay between learning attempts.
      - *Long ITIs*: May span 5–15 minutes, depending on whether memory consolidation or fatigue effects are being investigated.

        The choice of interval can influence learning dynamics and should be matched to the specific objectives of the study.
    - Response to Failed Platform Location

      In cases where an animal does not locate the hidden platform within the allotted trial duration, usually 60–90 seconds, corrective measures are employed to facilitate learning:
      - *Passive Placement*: The animal is gently placed on the hidden platform and held there for approximately 15 seconds, enabling it to associate the platform's location with distal visual cues.
      - *Guided Placement*: In some protocols, animals are gently guided to the platform, rather than placed passively. This method may enhance spatial encoding by involving active navigation.

- **Apparatus Design and Setup**

  The MWM consists of a large, open, and circular pool filled to approximately half its depth with water. The design intentionally minimizes distinct features within the pool itself, creating a relatively featureless interior. This setup reduces the availability of proximal cues, thereby encouraging the animal to rely on distal environmental cues for spatial navigation and learning.

- Species-Specific Apparatus Dimensions
  - *Rat Maze*: A circular stainless-steel tank measuring 210 cm in diameter with side walls 51 cm high.
  - *Mouse Maze*: A smaller version measuring 122 cm in diameter, also with 51 cm high side walls.

    Both maze types are typically elevated above the floor and equipped with plumbing systems, including drain lines and valves, to facilitate routine cleaning and water management.

- Color Adjustments for Strain and Visibility

  Visual contrast is an important factor in water maze performance, especially for albino animals:
  - The rat maze used with albino strains is painted black on the inside to enhance visual cue detection.
  - The mouse maze, typically used with pigmented strains like C57BL/6, is painted white to provide optimal contrast with the animal's coat color.

- Platform Shape, Size, and Positioning

  The platform may be either circular or square in shape. Typical dimensions include:
  - *Area-Based Platforms*: approximately 10–11 cm^2
  - *Diameter-Based Platforms*: commonly 10–12 cm in diameter

    Platforms are generally submerged 1–2 cm below the water's surface in experiments involving rats, while in mice, the depth is often reduced to 0.5–1 cm to better accommodate their smaller size and swimming capabilities.

- Platform Materials and Construction

  Platforms are commonly constructed from acrylic or polyvinyl chloride (PVC), offering durability and stability. The most prevalent configurations include:
  - A vertical post, often made from acrylic dowel or PVC pipe, with the platform mounted on top.
  - A textured surface (e.g., small holes or roughened finish) to provide traction for the animal.
  - A broad acrylic base that rests on the pool floor, designed to prevent tipping and minimize movement if bumped.

- Water Temperature

  Another important consideration in the MWM test is maintaining the water temperature within the optimal range of 19–22 °C. Deviations above or below this range may pose risks of thermal stress or hypothermia to the animals, potentially compromising both welfare and data integrity.

- Room Lighting Conditions

  Proper room lighting is essential in MWM experiments, particularly when using video tracking systems and computer-based software to monitor animal performance.

  To ensure optimal recording conditions:

  Lighting should be indirect to prevent glare and reduce surface reflections.

  Direct lighting can create unwanted reflections on the water, which may interfere with the accuracy and consistency of tracking software.

- Experimenter Presence and Positioning

  The position of the experimenter within the testing room can significantly influence behavioral outcomes in the MWM. Animals may use the experimenter as a distal spatial cue, which can affect navigation and latency, particularly when swimming toward or away from the experimenter's location. To reduce potential confounding effects, the experimenter may adopt one of the following approaches:

- Remaining stationary in a fixed location (the most commonly used approach)
- Standing behind a barrier to obscure visibility
- Leaving the room entirely during trials

- Establishing Consistent Spatial Coordinates

  To establish a consistent spatial reference system within the MWM, two imaginary perpendicular lines are drawn across the diameter of the circular pool, forming a "+" configuration that divides the arena into four equal quadrants.

  The experimenter must clearly label each quadrant using a consistent and standardized reference framework. These labels are used uniformly across all subjects and sessions to accurately describe the location of:

  - The hidden platform
  - Start positions
  - Quadrant occupancy during behavioral analysis

    However, it is important to note that the traditional four-quadrant setup may not be appropriate for certain animal models that display persistent thigmotaxis behavior despite repeated training sessions. In experimental settings, thigmotaxis is commonly regarded as a behavioral marker of anxiety and stress, with rodents exhibiting heightened anxiety often showing a preference for remaining close to the pool walls (8).

- **Limitations and Disadvantages**
  - Stress-Induced Confounds

    The MWM is inherently stressful because rodents are forced to swim in water, an aversive condition. This stress elevates corticosterone levels in the brain and blood plasma, potentially affecting performance outcomes. The relationship between stress and learning in MWM follows an inverted U-shaped curve in which both too little and excessive stress impair learning. Thus, careful optimization of trial number and training duration is essential to minimize confounding effects from stress.

  - Visual Acuity and Non-Cognitive Factors

    Visual function is a critical factor that can confound the interpretation of spatial learning performance in the MWM. Animals with intact visual acuity can detect and navigate toward the hidden platform using distal visual cues, contributing to better spatial learning. Conversely, animals with impaired vision, due to conditions such as eye enucleation, binocular deprivation, or age-related photoreceptor degeneration, exhibit poorer performance in spatial tasks.

    To control for the influence of vision and other non-cognitive factors, the cued (visible platform) version of the water maze is used. This test helps distinguish whether poor performance in the spatial version is due to cognitive deficits or sensory/motor impairments. The cued test can be administered either before or after the spatial learning assessment to validate the animals' ability to perceive visual cues (6) (Figure 8.1).

8.2.1.2 T-Maze Test

The T-maze test is a behavioral assay primarily used to study spatial memory and decision-making in rodents. Due to its simplicity, adaptability, and sensitivity to hippocampal integrity, it has become a fundamental tool in behavioral neuroscience, particularly in studies of cognitive dysfunction in disease models (9,10).

- **Apparatus and Design**

  The apparatus consists of a T-shaped maze comprising a start arm (stem) and two goal arms (left and right) extending at 180° angles. The maze is typically constructed from opaque

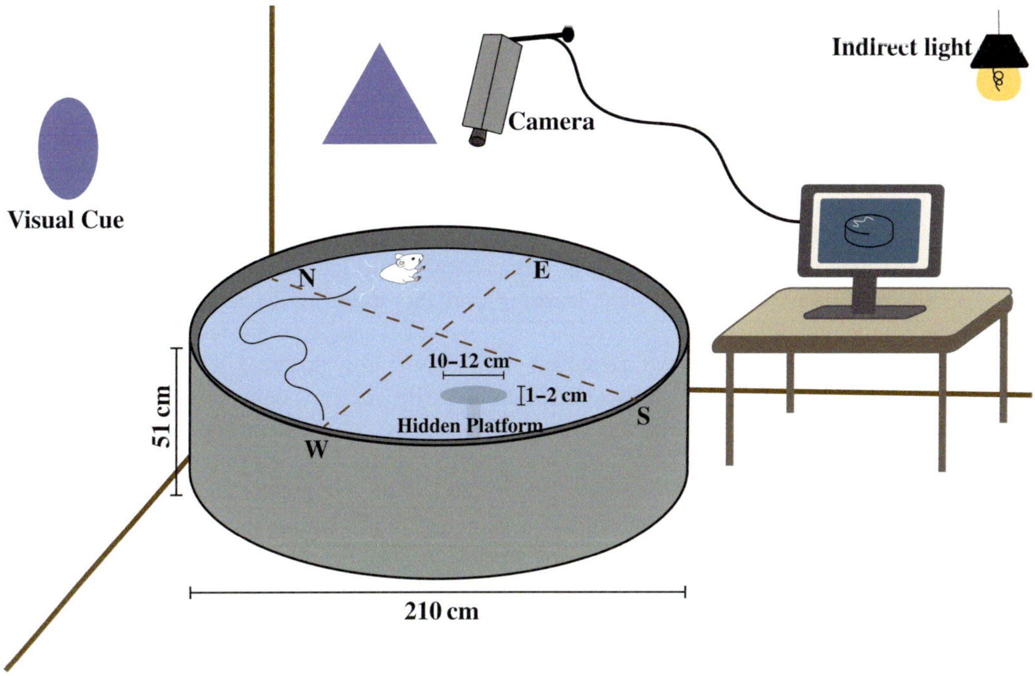

FIGURE 8.1 Schematic representation of the Morris water maze apparatus. (Photograph by the authors.)

acrylic or plastic, with surrounding walls to prevent escapes and minimize external distractions. Depending on the sophistication of the experimental setup, some T-mazes are equipped with automated features such as guillotine doors, food pellet dispensers, and infrared sensors to facilitate precise tracking and minimize human interference. While dimensions can vary slightly across laboratories, arms are generally 30–50 cm in length and 10–15 cm in width (11).

- **Historical Background and Concept**

  The T-maze originated from early behavioral research in the 1910s by Robert Yerkes, who first employed maze-like designs to study learning in invertebrates (12). Edward Tolman later adapted this paradigm in the 1920s to investigate cognitive processes in rodents, introducing the concept of spontaneous alternation, the natural tendency of rodents to explore novel environments by alternating arm choices (13). This concept was further refined by Wayne Dennis in the 1930s, who identified alternation as an expression of innate exploratory drive and memory of prior choices (14). Spontaneous alternation does not require prior training or reinforcement, making it ideal for quickly assessing working memory without the confounding effects of learning or motivation. The behavior relies on the rodent's capacity to retain short-term spatial information, a function heavily dependent on the hippocampus.

- **Standard Testing Procedure**

  In the typical spontaneous alternation protocol, rodents are placed at the start arm and allowed to choose freely between the two goal arms. Upon full entry into one arm, the animal is confined there for a brief period (usually ~30 seconds), then returned to the start arm for a subsequent trial. Testing sessions usually comprise 5–12 trials, with ITIs ranging from immediate succession to delays of up to 60 seconds. Performance is quantified as the percentage of trials in which the animal alternates its choice relative to the previous trial. Healthy rodents typically exhibit alternation rates between 70 and 75%, significantly above chance levels. The length of the ITI is a critical parameter: shorter intervals support better performance, whereas longer delays tend to impair alternation, reflecting the temporal limitations of spatial working memory. Like the MWM, the T-maze is sensitive to hippocampal damage, but it is particularly well-suited for detecting partial or subtle impairments in hippocampal function (9).

- **Variants and Experimental Considerations**

  Several variants of the T-maze test have been developed, each emphasizing different cognitive processes. The spontaneous alternation version is time-efficient, stress-minimizing, and does not rely on food motivation, making it especially suitable for quick cognitive screening. In contrast, the rewarded alternation version requires food restriction and extended habituation, spanning several days. While more labor-intensive, the rewarded paradigm allows for reinforcement learning and the study of decision-making under reward-based conditions.

  Interestingly, cognitive impairments are often more readily detected in the spontaneous version when short ITIs (e.g., 30 seconds) are used. By contrast, deficits in the rewarded version may only emerge at longer ITIs (e.g., 180 seconds), which place higher demands on memory retention (15,16).

  Another methodological distinction lies in the use of discrete versus continuous trials. In discrete trial formats, the animal is returned to the start arm after each choice, allowing for controlled trial timing and manipulation of ITIs. This format enhances the sensitivity of the task to detect working memory deficits. Conversely, continuous trials permit the animal to explore freely for a fixed duration or number of choices. Although less stressful due to reduced handling, continuous protocols may introduce variability that masks subtle cognitive impairments, particularly those dependent on hippocampal integrity (17,18).

  The physical design of the maze also plays a critical role in behavioral outcomes. Rodents typically exhibit less anxiety and greater exploratory behavior in enclosed mazes, which provide visual barriers and minimize exposure to open spaces. Enclosed T-mazes are thus preferred in cognitive studies, as they reduce the influence of anxiety and allow for cleaner assessments of memory and decision-making. In contrast, open mazes may elicit anxiety-like behaviors, such as thigmotaxis or freezing, which can confound cognitive measurements and often necessitate additional habituation sessions, thereby prolonging the testing schedule (19).

- **Extended Applications of the T-Maze**
  - Decision-Making Paradigms

    Beyond memory assessment, the T-maze is also utilized in decision-making studies. In these paradigms, one goal arm may offer a larger or more preferred reward that comes with a higher cost (e.g., delay, effort), while the other arm presents a smaller but more immediate reward. Such tasks enable researchers to investigate how animals evaluate trade-offs between reward magnitude and effort or time costs, providing insights into motivation, impulsivity, and executive function (20).
  - Novelty Preference and Short-Term Habituation

    The T-maze (and its three-arm variant, the Y-maze) can also be adapted to measure short-term habituation and novelty preference. In these designs, animals are first allowed to explore two of the three arms (with one arm blocked) during a training session. After a retention delay, all arms are opened for a test session. A preference for the previously inaccessible, novel arm reflects intact short-term spatial memory and exploratory drive. This paradigm is particularly useful for identifying memory impairments without the need for reinforcement or aversive stimuli (20).

- **Outcome Measures**

  The primary outcome metric in T-maze testing is the percentage of spontaneous alternation, defined as the proportion of trials in which the animal chooses an arm opposite to the one visited previously. A spontaneous alternation rate significantly above chance,

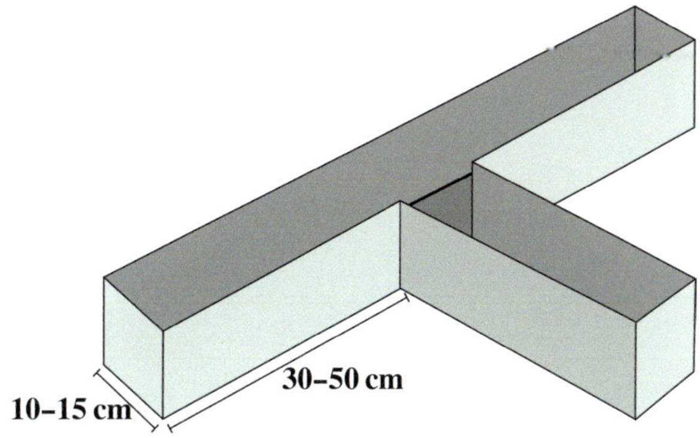

FIGURE 8.2 Schematic representation of the T-maze apparatus. (Photograph by the authors.)

typically ranging between 70 and 80%, is indicative of intact spatial working memory. Secondary measures such as arm entry latency and side preference are also recorded to control for motivational or anxiety-related confounds (Figure 8.2).

8.2.1.3 Y-Maze Test

The Y-maze test is a widely utilized behavioral paradigm designed to evaluate spatial working memory and spontaneous alternation behavior in rodents, particularly mice and rats. This task exploits the animal's innate tendency to explore novel environments and serves as a robust tool for assessing hippocampal function, cognitive flexibility, and memory-related impairments associated with various neurological and neuropsychiatric conditions.

- **Apparatus and Design**

  The Y-maze consists of three arms arranged at 120° angles, forming a Y-shaped configuration. Each arm is identical in appearance to avoid bias in exploration. For mice, typical arm dimensions are approximately 35 cm in length, 5 cm in width, and 20 cm in height.

  For rats, larger dimensions are employed, generally around 50 cm in length, 16 cm in width, and 30 cm in height. The maze is commonly constructed from opaque or transparent acrylic or polyvinyl material with a smooth, non-reflective, and non-slip floor surface to ensure safety and optimal locomotor activity. The design facilitates natural exploratory behavior, allowing rodents to freely traverse all three arms. This configuration is particularly suited for investigating spatial working memory, with specific relevance to hippocampal integrity. The Y-maze is frequently employed in preclinical research to study aging, neurodegeneration, pharmacological interventions, and genetic models of cognitive disorders (21,22).
- **Testing Procedure**

  Two main protocols are commonly used with the Y-maze: the spontaneous alternation test and the novel arm recognition test, each probing distinct aspects of cognitive performance.
  - Spontaneous Alternation Test

    In this protocol, the animal is placed either at the center or at the end of one arm and allowed to freely explore the entire maze for a fixed duration, typically 8 minutes. Rodents naturally exhibit a preference for entering a less recently visited arm, demonstrating a behavior known as spontaneous alternation. The sequence and number of arm entries are recorded to calculate the percentage of alternation, defined as successive entries into each of the three arms without repetition. This measure is considered a reliable index of spatial working memory and is sensitive to disruptions in hippocampal circuitry and other memory-related brain regions.

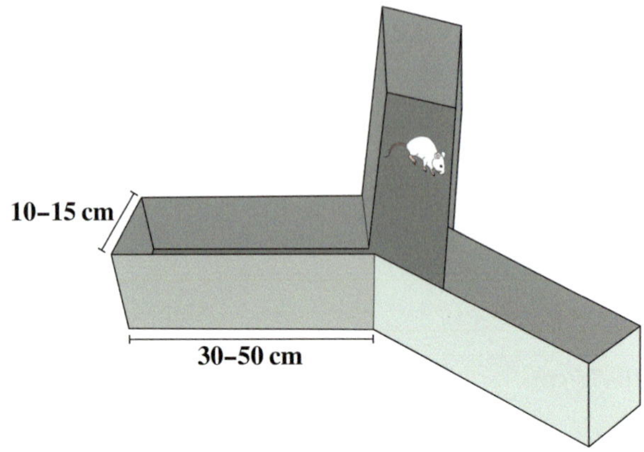

FIGURE 8.3 Schematic representation of the Y-maze apparatus. (Photograph by the authors.)

- Novel Arm Recognition Test (Two-Trial Y-Maze)
  This protocol involves two sessions. In the first session (training phase), one of the three arms is blocked off, and the animal is allowed to explore the remaining two arms for approximately 15 minutes. After a retention interval, the second session (test phase) is conducted, during which all three arms are opened. The animal is reintroduced into the maze, and its exploratory behavior is monitored. Rodents with intact spatial recognition memory typically show a preference for the previously blocked (novel) arm, as indicated by the proportion of time spent and number of entries into that arm. A preference ratio of 60–70% for the novel arm is generally interpreted as evidence of preserved short-term spatial memory (23,24).
- Data Collection and Interpretation
  Key behavioral metrics recorded during Y-maze testing include:
  - The total number of arm entries
  - The sequence of arm visits to assess alternation behavior
  - The percentage of correct spontaneous alternations
  - Time spent in each arm (especially the novel arm in recognition tasks)
  - The first arm entered during the test phase

Together, these parameters provide a comprehensive assessment of spatial memory performance, offering insights into hippocampus-dependent cognitive processing and the modulatory effects of pharmacological, genetic, or environmental manipulations (Figure 8.3).

8.2.1.4 Barnes-Maze Test

The Barnes maze test is a widely used behavioral assay to assess spatial learning and memory in rodents, especially mice and rats. It leverages rodents' natural aversion to open spaces and their preference for dark, sheltered areas to motivate them to find an escape box hidden beneath one of multiple holes arranged around a circular platform. Originally developed by Carol Barnes for use in rats (25), the paradigm has since been adapted for mice and has become a staple in preclinical neuroscience for assessing hippocampal-dependent memory processing (26). It resembles the MWM in that it requires hippocampal function for learning the relationship between environmental cues and a specific escape location.

- **Apparatus and Design**

The apparatus for this maze experiment features a circular elevated platform with a diameter of approximately 1 m, equipped with 12–40 evenly spaced holes (typically around 20), each measuring about 5 cm in diameter to accommodate rodents. Among these holes, one leads to a darkened escape box or home cage situated beneath the platform, while the others are false openings that lead to open spaces or dead ends. To aid in spatial orientation and learning, distal visual cues such as colored shapes or patterns are strategically placed around the testing room. The maze environment is brightly lit at around 1,000 lux to create an aversive atmosphere, motivating the rodents to seek refuge in the dark escape box, with white noise potentially masking background sounds. A video tracking system mounted overhead is used to record behavioral parameters throughout each trial (26,27).

- **Testing Procedure**

Each session begins with the animal placed under a starting cylinder in the center of the maze for an acclimation period of approximately 30 seconds. Each trial is designed to last up to 2 minutes, during which the animal is allowed to explore and locate the escape box; if it does not succeed within this timeframe, it is gently guided to the escape hole and permitted to rest there for 10–15 seconds. Typically, four trials are conducted over four consecutive days to evaluate the rodent's learning capabilities, followed by a probe trial on the fifth day in which the escape box is removed to assess memory retention. An optional reversal learning phase can be introduced, where the escape box location is changed to evaluate cognitive flexibility and the ability to acquire new memories. To maintain the integrity of the experiment, the maze is thoroughly cleaned with 70% ethanol between trials or animals to eliminate olfactory cues. Key parameters measured during the trials include latency to find the escape hole, number of errors made by visiting incorrect holes, path length and search strategy, number of target zone crossings during the probe trial, as well as the velocity and exploration patterns of the rodent (26,28) (Figure 8.4).

8.2.2 Object Location Test

The object location Test (OLT) is a straightforward and reliable behavioral assay used to evaluate hippocampus-dependent spatial memory. It capitalizes on the rodent's innate tendency to explore novel spatial configurations, without requiring external rewards or punishments. As a result, the OLT minimizes potential confounding effects related to stress or emotional reactivity. These features make it a low-burden and ethologically valid method for assessing spatial memory in rodents (29).

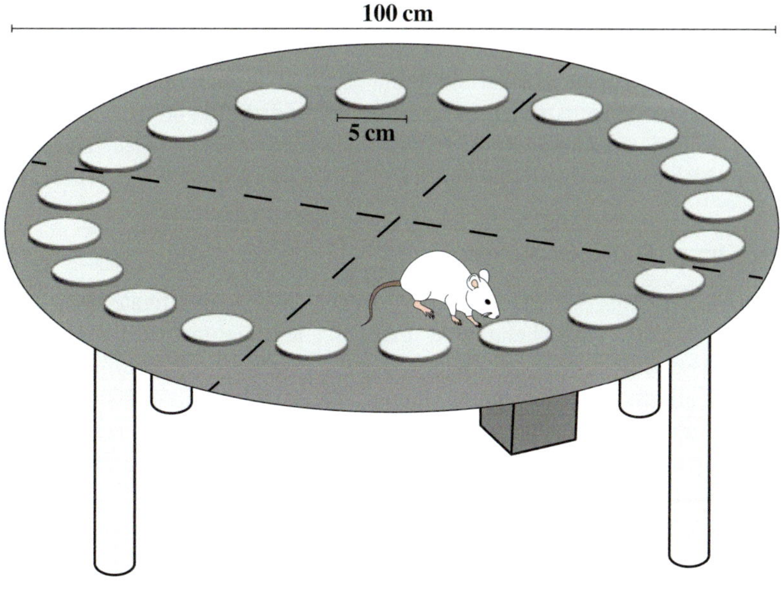

FIGURE 8.4 Schematic representation of the Barnes maze apparatus. (Photograph by the authors.)

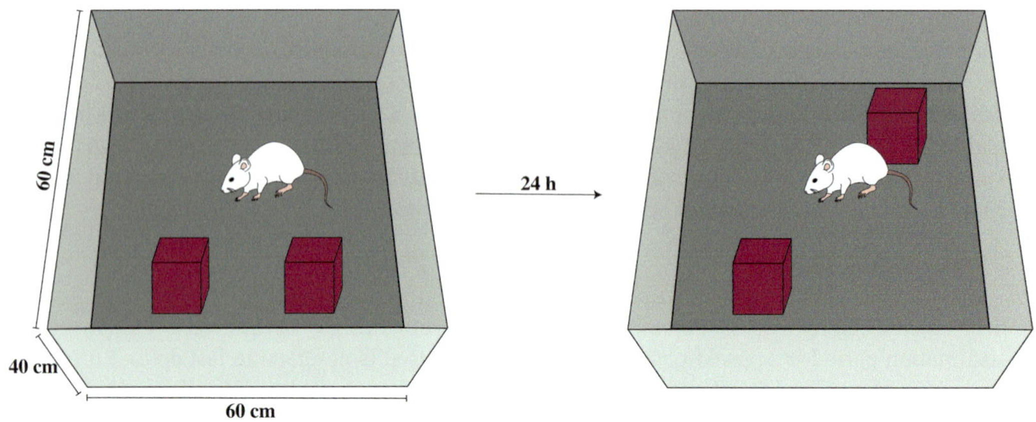

FIGURE 8.5 Schematic representation of the object location test apparatus. (Photograph by the authors.)

- **Testing Procedure**

Initially, each mouse undergoes habituation to an open-field testing arena, allowing it to become familiar with the environment. During the sample phase, the mouse is exposed to two identical objects, placed in specific locations within the arena, and is allowed to explore freely while using spatial cues from the environment. Following this initial exploration, the mouse is temporarily removed from the arena for a delay period. During this time, one of the objects is relocated to a new position within the arena. The mouse is then returned for the test phase, where it can again explore both objects. Under normal conditions, mice exhibit a natural preference for novelty, and those with intact spatial memory are expected to spend more time investigating the object in the novel location. In contrast, animals with hippocampal damage, which impairs spatial contextual learning, typically show no clear preference between the familiar and relocated object, reflecting deficits in spatial memory encoding or retrieval (30) (Figure 8.5).

8.3 AMNESIA

One of the most compelling areas of research in neuroscience involves exploring how novel experiences influence behavior in both animals and humans. Novelty can be defined as a deviation from what is expected, based on prior experiences and internal predictions about the likelihood of certain outcomes. When animals encounter novel stimuli, these experiences often trigger physiological stress responses, which may lead to behavioral changes such as approach behavior (31).

8.3.1 Novel Object Recognition Test

In 1988, Ennaceur and Delacour were the first to develop and investigate the NOR and Novel Location Recognition tests. These behavioral tests offer a straightforward and ethologically relevant approach to assessing memory in rodents. They capitalize on the animal's innate tendency to explore novel stimuli, thereby eliminating the need for external reinforcements such as rewards or punishments. Although the NOR test was initially developed to assess recognition memory, it can be adapted to investigate other cognitive and behavioral domains, including working memory, attention, anxiety-related behavior, and preference for novelty (32,33).

Researchers employ a range of scoring systems in the NOR test, depending on the specific objectives of their study. Commonly used measures include the Discrimination Index, which quantifies the animal's ability to distinguish between novel and familiar objects; the Global Habituation Index, which reflects the overall decrease in exploratory behavior over time; and the Preference Index, which indicates the relative interest in the novel object compared to the familiar one (34,35).

- **Testing Procedure**

The NOR test consists of three main phases:

1. Habituation Phase
During the habituation phase, the animal is introduced into an open-field arena that contains no objects. It is allowed to freely explore the environment for a defined period, enabling it to become familiar with the context. Following this period of exploration, the animal is returned to its home or holding cage. This phase serves to minimize anxiety or stress responses associated with exposure to a novel environment, thereby reducing potential confounding effects during subsequent testing phases.
2. Familiarization Phase
In this phase, the animal is reintroduced into the open-field arena, which now contains two identical objects (e.g., A and A) positioned in opposite corners. The animal is allowed to explore freely for a few minutes, providing the opportunity to investigate and encode the features of the objects. To prevent initial bias and ensure voluntary exploration, the animal is placed in the arena facing the wall opposite the objects at the start of the session.
3. Test Phase
Following a retention interval, short or long, depending on the experimental design, the animal is once again placed in the open-field arena. This time, the arena contains one familiar object (A) and one novel object (B), arranged in an A–B configuration. To minimize positional bias, the placement of the novel and familiar objects is counterbalanced across subjects and trials. Both objects are positioned in symmetrical, opposite corners of the arena, maintaining consistency with the familiarization phase and allowing for reliable measurement of novelty preference (36).

- **Expected Behavior and Interpretation**

Memory performance is quantified using a preference score, typically expressed as either a percentage of total exploration time spent on the novel object or as a discrimination index (e.g., the difference in time spent exploring the novel versus the familiar object divided by total exploration time). Under normal conditions, healthy animals typically display a strong preference for the novel object, spending significantly more time exploring it during the initial minutes of the test phase. This behavior indicates that the animal recognizes the familiar object and, therefore, demonstrates intact recognition memory. Importantly, recognition performance can be influenced by repeated exposure. When animals are repeatedly familiarized with the sample object over several days, they are capable of retaining recognition memory for extended periods, even up to several weeks (37).

Additionally, novelty preference is time-sensitive. The strongest discrimination between novel and familiar objects typically occurs early in the test phase, when the novel object is still perceived as unfamiliar. As the test progresses, the novelty of the object diminishes, leading to a reduction in exploration time and, consequently, a decline in preference bias (36).

Over years of widespread use, the general protocol for the NOR test has become largely standardized. However, one critical aspect remains inconsistent across studies: the selection and implementation of the objects themselves. A key consideration is determining whether an animal's exploratory behavior is driven by the novelty of the object or by an intrinsic preference for specific object features. Research has shown that certain objects are consistently approached or avoided by animals, regardless of their novelty status. Moreover, an object pair that elicits a strong novelty preference in one

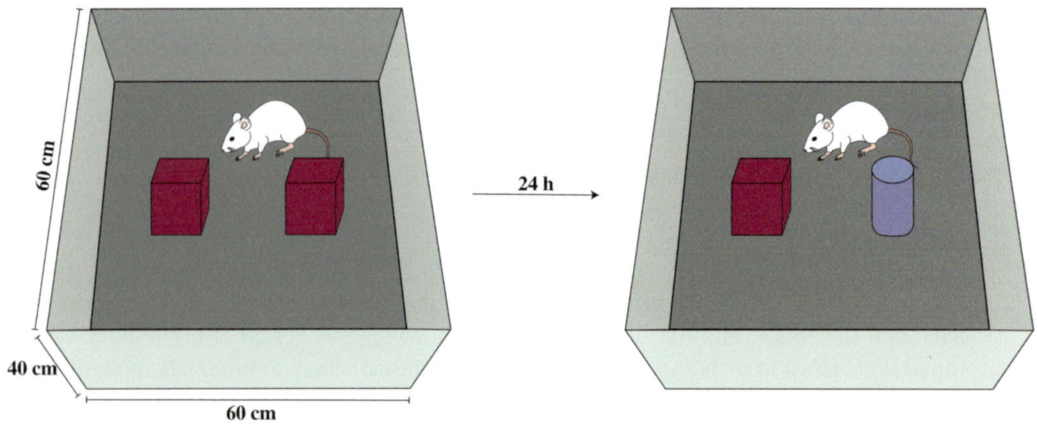

FIGURE 8.6 Schematic representation of the novel object recognition test apparatus. (Photograph by the authors.)

group of animals may not produce the same outcome in another. These findings highlight the necessity of employing counterbalanced designs, in which object placement and assignment are systematically varied to minimize bias and enhance the reliability of the task. Detailed reporting of object characteristics, consistent use of counterbalancing, and careful differentiation between novelty-induced anxiety and recognition memory are essential for improving reproducibility and ensuring the scientific validity of results obtained from the NOR test (38).

- **Equipment**

  Open-field arenas used in this test exhibit considerable variability in their physical characteristics, including size, shape, and color. The shape of the arena is an important design element and can influence the interpretation of exploratory behavior. Common shapes include square, circular, and rectangular configurations. However, the open-field arenas in most commercially available systems are typically square or rectangular.

  Despite this lack of direct evidence, relevant behavioral data suggest that arena shape may not significantly influence general activity patterns (39) (Figure 8.6).

  Overall, while the NOR task remains a widely used and valuable tool for assessing recognition memory in rodents, emerging evidence underscores the need to interpret novelty preference within a broader behavioral context, rather than as a direct proxy for memory performance alone.

8.4 ANXIETY-LIKE BEHAVIOR TESTS

Anxiety is a mental illness characterized by cognitive, emotional, behavioral, and physical symptoms, derived from the Latin word "Ango," meaning "to vex or torment." It can evoke feelings of worry, fear, and unease, and is sometimes a normal response to stress 666.

8.4.1 Open-Field Test

The open-field test, originally developed in the 1930s by Hall and Ballachey, remains one of the foundational behavioral assays used to evaluate anxiety-like behavior, exploration, and locomotor activity in rodent models. While it was initially introduced as a method to assess rodent timidity based on defecation frequency (40), the test has since evolved into a versatile paradigm for measuring a wide range of behavioral traits, including fearfulness, exploration, and anxiety-like behavior such as rearing and thigmotaxis (41).

Behaviorally, the open field test reflects a conflict between the rodent's exploratory drive and defensive drive. The novel, large open arena is inherently stressful because it separates the animal from its home environment and exposes it to a potentially threatening space. Rodents often exhibit thigmotaxis, preferring to stay close to the walls, which is considered an indicator of anxiety-like behavior rather than a direct fear response to a specific stimulus 668.

- **Apparatus Design and Setup**
  The testing apparatus consists of a square or rectangular arena, typically enclosed by opaque walls to prevent escape. Common dimensions range from 40 × 40 × 30 cm to 100 × 100 × 50 cm, depending on the species and experimental design. The floor is often marked with a grid or divided into defined zones, including center, periphery, and corners, to facilitate spatial analysis. The arena is monitored by overhead video cameras, and behavior is captured using automated tracking software such as EthoVision, SMART Video Tracking, Autovision, or Accuscan Versamax.
- **Testing Procedure**
  To minimize stress and ensure reliable behavioral outcomes, animals are acclimated to the testing room for at least 30 minutes prior to the onset of behavioral assessments. During the test, each mouse is gently placed in the center of the open-field arena and allowed to explore freely for a period of 5–10 minutes, while behavior is continuously recorded using video tracking systems.
  Key behavioral parameters assessed include:
- *Locomotor Activity*: total distance traveled, time spent moving
- *Anxiety-Like Behaviors*: time spent in the center versus periphery, number of center entries
- *Exploratory Behaviors*: frequency of rearing and stretch-attend postures

Behavioral data are typically analyzed according to spatial zones within the arena (e.g., center, periphery, corners), providing insight into anxiety and exploratory tendencies. Upon completion of each session, the animal is returned to its home cage. The arena is then thoroughly cleaned with disinfectant to eliminate olfactory cues that may affect subsequent subjects. To avoid unintended behavioral influence, human presence during testing should be minimized or visually concealed (42) (Figure 8.7).

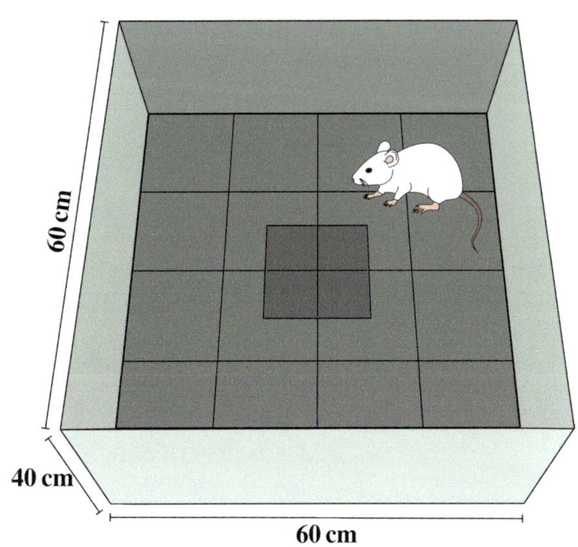

FIGURE 8.7 Schematic representation of the open-field test apparatus. (Photograph by the authors.)

8.4.2 Hole-Board Test

The hole-board test, originally developed by Boissier et al., is a well-established behavioral assay used to evaluate neophilia, exploratory drive, and anxiety-like behaviors in rodents. It merges the approach-avoidance conflict seen in open-field paradigms with added environmental complexity, floor holes, to more sensitively detect subtle changes in exploratory behavior and emotional reactivity (43,44).

- **Apparatus Design and Setup**

The apparatus is typically a square or rectangular arena with multiple holes (ranging from 4 to 36) evenly spaced on the floor. The size and number of holes vary depending on whether rats or mice are tested. Behavior can be recorded using video tracking systems (e.g., Noldus EthoVision® XT) or live-scored by trained observers.

- **Testing Procedure**

Each animal is placed at a designated start corner facing the hole board. The test lasts 5 minutes per animal.

Among the main behaviors is the "head-dip" behavior, where the rodent inserts its head into a hole. The frequency, duration, and latency of head-dips are recorded. Head-dipping is considered an indicator of neophilia (attraction to novelty) and inversely related to anxiety; the more head-dips, the lower the anxiety level. Other behaviors such as sniffing, immobile sniffing, climbing, rearing, various grooming actions, and locomotion are also observed to provide a fuller picture of the animal's anxiety state and exploratory behavior (45) (Figure 8.8).

8.4.3 Contextual and Cued Fear Conditioning Test

Fear conditioning is a widely employed behavioral paradigm in rodent studies to evaluate associative learning and memory related to aversive experiences. This test assesses the animal's capacity to form and retain associations between an aversive unconditioned stimulus, typically a mild electric foot shock, and specific environmental cues. These cues may include the overall context in which

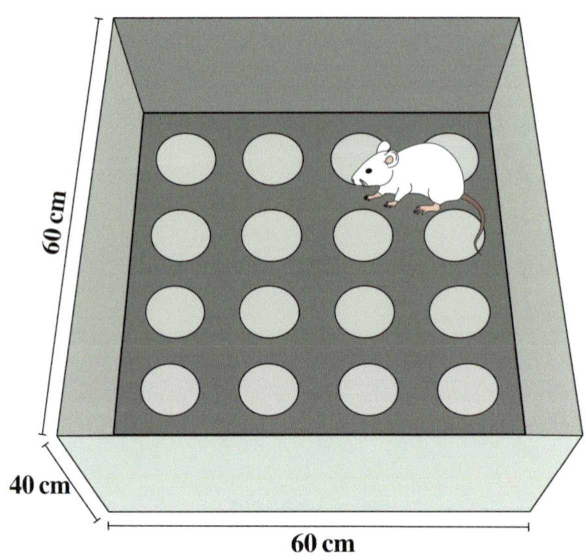

FIGURE 8.8 Schematic representation of the hole-board test apparatus. (Photograph by the authors.)

the shock occurs (contextual conditioning) or a discrete signal, such as a tone, that precedes the shock (cued conditioning) (46).

- **Experimental Design and Setup**

  The procedure is conducted in a specialized conditioning chamber, typically a transparent acrylic box with a metal grid floor for delivering foot shocks. An integrated speaker allows for the presentation of auditory cues, such as a tone. The chamber is usually placed in a sound-attenuated room to minimize external noise interference. Environmental factors, including lighting, scent, and floor texture, are carefully controlled to ensure consistency across experimental sessions. This setup enables precise evaluation of the animal's capacity to form fear-related associative memories based on either environmental or discrete sensory cues (47).
- **Testing Procedure**

  The preparation and setup for fear conditioning experiments require a controlled and quiet environment to minimize external stimuli. Utilizing a sound-insulated anteroom to house the mice helps prevent any exposure to auditory cues that could influence the conditioning process. During contextual conditioning, the animal is placed in a novel chamber, where it is allowed to habituate for 120 seconds before a tone cue (at 70–80 dB) is presented for 15–30 seconds, culminating in a mild foot shock (0.6 mA) administered during the last 2 seconds of the tone. Following an inter ITI of 60–210 seconds, the tone-shock pairing is repeated, and after the final trial, the animal should remain in the chamber for an additional 60 seconds before being removed.

  For context testing, it is crucial to conduct the assessment in the same environment used during training to maximize conditioning effects. The animal should be placed in the chamber for 3–5 minutes without the tone cue, with observations made to note any freezing behavior, and it should be removed promptly after testing. If cue testing is to be performed, the conditioning context must be thoroughly cleaned or a novel environment should be utilized, allowing for 60 minutes of habituation before presenting the tone cue for 3 minutes, followed by an additional minute of observation without the cue. Trace conditioning follows a similar protocol as delay conditioning but incorporates a trace interval of 2–60 seconds between the end of the conditioned stimulus and the onset of the unconditioned stimulus, typically involving five trials to ensure adequate associative learning. Freezing behavior should be recorded during both contextual and cued testing for subsequent analysis. Data analysis will focus on the duration of freezing as an indicator of fear response (47).

  Key brain regions implicated in contextual and cued fear conditioning include the amygdala, hippocampus, and frontal/cingulate cortex. The amygdala plays a critical role in processing fear responses, with specific nuclei contributing to both contextual and cued conditioning. In contrast, the hippocampus is crucial for contextual learning but is less involved in cue learning. The frontal and cingulate cortices are associated with attentional processes and memory acquisition, and lesions in these areas can impair contextual conditioning. Understanding the functional contributions of these brain regions is essential for dissecting the neural mechanisms underlying fear learning and memory (48,49) (Figure 8.9).

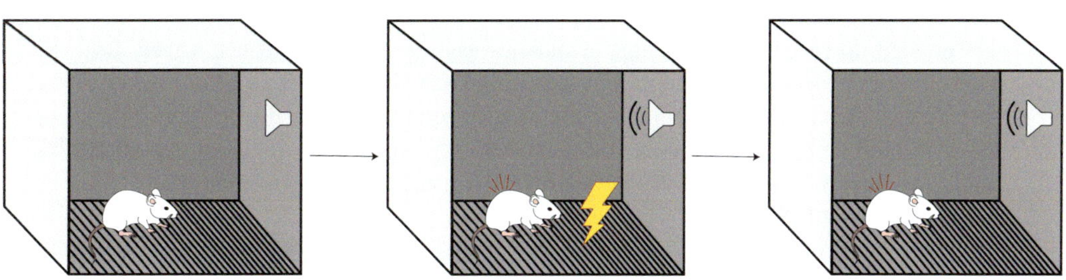

FIGURE 8.9 Schematic representation of the fear conditioning test apparatus. (Photograph by the authors.)

8.4.4 Passive Avoidance Test

The passive avoidance test is a well-established behavioral paradigm designed to assess associative learning and memory in rodents, particularly long-term memory and emotional learning. This test typically includes two primary variants: the step-through and step-down inhibitory avoidance tasks (IATs) (50). Early experimental models of learning and memory date back to Thorndike's "problem boxes," where rodents learned to perform specific actions, such as pressing a lever, to obtain food. In the late 1930s, Hudson and colleagues introduced the IAT, later refined by Jarvik and Kopp in the 1960s to incorporate aversive conditioning using a mild foot shock in a two-compartment setup. This refinement led to the development of the step-down inhibitory (passive) avoidance task, which remains widely used today (51).

The passive avoidance task engages key brain structures involved in the integration of contextual cues and emotional responses. The hippocampus and amygdala are central to this network, playing essential roles in the formation and retrieval of aversive memories. Neurotransmitter systems, including dopamine and substance P, further modulate these processes, influencing both acquisition and expression of avoidance behavior (52,53).

1. Step-Through Passive Avoidance Task

   The step-through avoidance test is a behavioral assay primarily used to evaluate learning and memory, especially long-term memory and emotional learning, in small laboratory animals such as rodents. It is based on negative reinforcement and fear motivation, where animals learn to avoid an environment previously associated with an aversive stimulus, typically a mild foot shock (54,55).

   - **Experimental Setup and Design**

     The apparatus consists of a brightly lit (white) compartment and a dark (black) compartment separated by a guillotine or sliding door. The compartments are sized appropriately for the species tested (mice or rats). The bright compartment is aversive due to illumination, while the dark compartment is preferred by rodents. The compartments are sized appropriately for the species tested (mice or rats). The dark compartment contains a grid floor capable of delivering a mild, controlled electric foot shock (55).

   - **Testing Procedure**

     During the acquisition (training) phase, the animal is placed in the illuminated compartment. After an initial exploration period, the door between the two chambers is opened. Upon entering the dark compartment, the door closes automatically, and a mild foot shock (0.2–0.5 mA for 1–2 seconds) is delivered. This conditions the animal to associate the dark environment with an aversive stimulus.

     To assess memory formation and retrieval, a retention test is conducted 24 hours later. The animal is reintroduced into the lit compartment with the door open. The latency to enter the dark compartment, a measure of memory retention, is recorded, with longer latencies indicating stronger aversive memory. All behavioral parameters can be recorded manually or via automated tracking systems for enhanced accuracy (56).

2. Step-Down Inhibitory Avoidance

   The step-down passive inhibitory avoidance task is a behavioral test widely used to assess aversive memory in rodents by measuring their ability to avoid a previously experienced foot shock. This task relies on the association between a specific environment (a platform) and an aversive stimulus (electric shock), with memory retention quantified by increased latency to step down from the platform during testing (51).

   - **Apparatus and Setup**

     The apparatus typically consists of a clear acrylic chamber with stainless steel grid bars (0.3–2 mm wide, spaced 1 cm apart) and an elevated platform (often in the corner of a chamber). Modern designs employ rectangular bars arranged in dipole pairs to invert

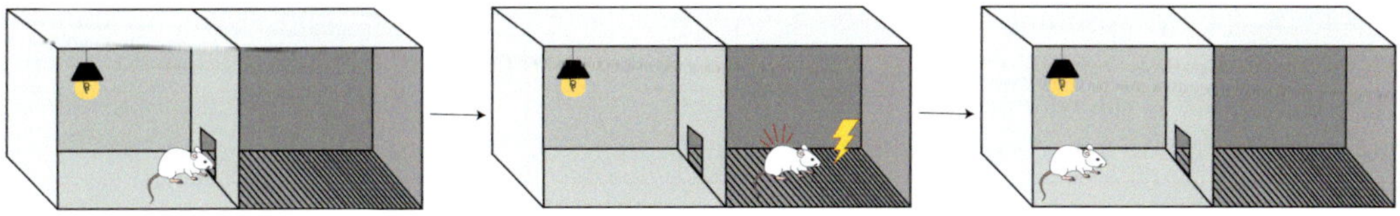

FIGURE 8.10 Schematic representation of the passive avoidance test apparatus. (Photograph by the authors.)

electrical polarity, ensuring consistent current delivery despite variations in the animal's bioimpedance. Scrambled foot shocks are administered via the grid. Automated systems track step-down latency (time to leave the platform) and escape latency (time to return) using software like Ethovision XT (51).

- **Testing Procedure**

  The protocol for the step-down passive inhibitory avoidance task consists of two main sessions: a training session and a test session, with a 24-hour interval between them. In the training session, the animal is gently placed on the platform facing a specific corner, such as the rear left corner. Once the animal steps down onto the grid with all four paws, a mild electric shock is delivered, commonly set between 0.35 and 0.5 mA at a frequency of 62–125 Hz for a duration of 2–3 seconds. Following the shock, the animal is immediately removed from the apparatus, and the time taken for the animal to step down, known as the step-down latency, is recorded, with a typical cutoff time of 60 seconds during this training phase (51,57).

  The test session occurs 24 hours later, where the animal is again placed back on the platform in the same manner as during training. This time, however, no shock is delivered when the animal steps down. The step-down latency is recorded again, with a usual cutoff of 180 seconds. After the animal either steps down or reaches the cutoff time, it is removed from the apparatus and returned to its home cage. An alternative protocol can be utilized in which, during the training session, the animal receives a continuous scrambled foot shock until it spontaneously returns to the platform; in this variant, the escape latency is recorded with a cutoff at 60 seconds. The subsequent test session remains the same, with the step-down latency recorded for up to 180 seconds (51,58) (Figure 8.10).

8.5 DEPRESSION TESTS

Depression is a multifaceted mood disorder characterized by both psychological and neurobiological disturbances. At the neural level, it arises from dysregulation within a complex network of brain regions involved in mood regulation, motivation, and cognitive function. Key structures implicated in this network include the prefrontal cortex, anterior cingulate cortex, amygdala, hippocampus, and striatum. Functional and structural abnormalities in these regions, such as reduced activity or volume, disrupted functional connectivity, and imbalances in key neurotransmitter systems, are consistently observed in individuals with depression. These alterations are thought to underlie core depressive symptoms, including anhedonia, cognitive impairments, emotional dysregulation, and diminished motivation (59,60).

8.5.1 Forced Swim Test

The forced swim test serves as a widely accepted preclinical model for studying depressive-like behavior and screening potential antidepressant drugs in rodents. The rationale behind the test is based on the observation that when an animal is placed in an inescapable water-filled tank, it first tries to escape by swimming and climbing. Over time, as escape appears impossible, the animal becomes immobile, floating passively with only minimal movements to keep its head above water. This immobility is interpreted as behavioral despair or learned helplessness, which are considered analogous to certain aspects of human depression (61,62).

- **Apparatus and Design**

The primary apparatus used in this test is a transparent cylindrical tank, commonly constructed from Plexiglas or glass. The tank typically measures between 20 and 30 cm in diameter and 30–60 cm in height. It is filled with water maintained at room temperature (approximately 24–25°C) to a depth that prevents the animal from touching the bottom or the sides with its paws or tail, thereby eliminating any possibility of escape. The water level is usually set at approximately two-thirds of the tank's height – for example, in a 60 cm tall tank, the water depth would be around 40 cm (62,63).

- **Testing Procedure**

The forced swim test procedure varies between rats and mice while maintaining a consistent focus on animal welfare. For both species, researchers must observe the animals continuously during the test and remove any that show signs of severe distress, such as an inability to stay afloat. Before testing, animals acclimate to the testing environment for 30–60 minutes to reduce stress caused by changes in their surroundings.

In the case of mice, the test consists of one session lasting 6 minutes, which includes a 2-minute pretest for acclimatization followed by a 4-minute testing phase.

For rats, the protocol involves two sessions spaced 24 hours apart. The first session serves as a 15-minute pretest, while the second session is a 5-minute test.

After each session, rats or mice are also placed in a drying cage with a heat source and monitored to maintain an appropriate recovery temperature. Researchers change the water after every session to prevent any influence on subsequent tests, ensuring a standardized approach for both species throughout the experiment (63,64).

Some studies indicate that the timing of experiments may affect the outcomes of the forced swim test. Therefore, it is advisable to conduct all experiments under consistent timing and conditions (62) (Figure 8.11).

FIGURE 8.11 Schematic representation of the forced swim test apparatus. (Photograph by the authors.)

8.5.2 Tail Suspension Test

The tail suspension test is a widely used behavioral assay in mice to evaluate depressive-like behavior and to screen potential antidepressant drugs.

- **Apparatus Design and Setup**

The primary component is a suspension bar or shelf ledge from which the mice or rats can be suspended by their tails. In addition, a tail suspension box is used, which is compartmentalized to prevent the mice from touching the walls or escaping. Each compartment is sized to approximately 35–55 cm in height, 15 cm in width, and 11.5 cm in depth, ensuring that the mouse's nose remains about 20–25 cm above the floor. A contrasting background color, such as dark grey for albino mice or cream for darker strains, enhances visibility during the test. The bottom of each compartment has a detachable tray to collect feces and urine, aiding in post-test cleanup. A stopwatch is necessary for timing immobility and mobility, enabling researchers to gather precise data regarding the mouse's behavior during the suspension period. A video camera can capture the complete view of the suspended animal, allowing for a thorough analysis of its behavior throughout the experiment (65).

- **Testing Procedure**

The animals should be allowed to acclimate to the testing room for at least 1 hour before testing. When preparing the animals for suspension, adhesive tape strips are cut. It is crucial to place the tape gently on the mouse's tail, ensuring minimal discomfort and securing the tail ends in the home cage to prevent tangling. The test lasts for 6 minutes, during which the behavior of the animal is observed and recorded. Immobility time is typically measured over the entire duration or specifically during the last 4 minutes, when initial escape attempts have subsided. After the test concludes, the tape must be gently removed from the tail to prevent injury or pain. The animal is then returned to its home cage, and the apparatus, along with the trays, should be thoroughly cleaned using the sterilizing solution. Behavioral scoring involves categorizing the actions into immobility and mobility, with immobility defined as passive hanging without escape-oriented movements. Mobility includes struggling, climbing, swinging, and other limb movements (66,67) (Figure 8.12).

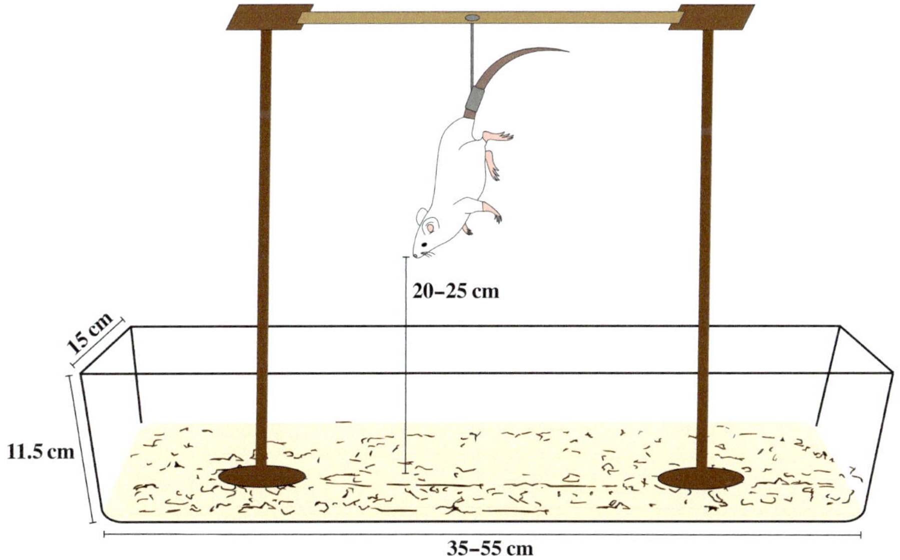

FIGURE 8.12 Schematic representation of the tail suspension test apparatus. (Photograph by the authors.)

8.5.3 Sucrose Preference Test

The sucrose preference test is a widely used behavioral assay to assess anhedonia, the reduced ability to experience pleasure, in rodent models, primarily mice and rats. It is based on a two-bottle choice paradigm where animals choose between a sweetened sucrose solution and plain water (68,69).

- **Apparatus Design and Setup**

Two identical bottles, one filled with a sucrose or saccharin solution (typically at a 1–2% concentration) and the other with plain drinking water, are placed in the animal's home cage.

- **Testing Procedure**

A systematic review of sucrose preference test protocols published in 2021 highlighted considerable variability in methodological approaches across studies, raising concerns about reproducibility and cross-study comparisons. The review identified significant differences in key procedural elements, including habituation practices, food and water deprivation protocols, test duration, and the type and concentration of the sucrose solution used. Notably, the analysis reported 65 distinct habituation protocols and 104 different combinations of deprivation and testing durations, emphasizing the need for greater standardization in behavioral assays assessing anhedonia (68).

Despite this variability, a widely adopted protocol involves an initial habituation phase, during which animals are exposed to two identical water bottles placed in their home cage for two to three days. This step minimizes the influence of side bias during the testing phase. Following habituation, animals are given access to two bottles, one containing a 1–2% sucrose (or saccharin) solution and the other containing plain drinking water, typically for a period of 12–24 hours. To control for potential side preference, bottle positions are switched midway through the test. Fluid consumption is determined by weighing the bottles before and after testing, and sucrose preference is calculated using the following formula:

$$\text{Sucrose Preference}\,(\%) = \left[\text{Sucrose intake}/\left(\text{water intake} + \text{sucrose intake}\right)\right] \times 100.$$

A reduced preference for sucrose is interpreted as an index of anhedonia. Throughout the procedure, animals should have ad libitum access to food, and environmental conditions such as lighting, temperature, and cage enrichment must be kept consistent to avoid confounding influences on behavior (69,70) (Figure 8.13).

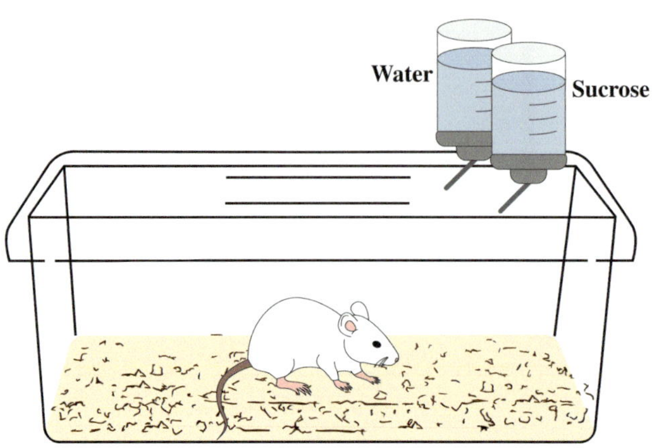

FIGURE 8.13 Schematic representation of the sucrose preference test apparatus. (Photograph by the authors.)

8.5.4 Elevated Plus Maze

The elevated plus maze (EPM) is a widely employed behavioral assay for evaluating anxiety-like behaviors in rodents. Originally introduced by Montgomery in 1955, the paradigm was significantly refined by Handley and Mithani in 1984 and later standardized by Pellow and colleagues in 1985, resulting in a maze with two open and two enclosed arms arranged in a plus (+) configuration. Subsequently, in 1987, Lister adapted the EPM specifically for use in mice, thereby broadening its applicability across rodent species (71).

- **Apparatus Design and Setup**

  The EPM consists of four arms, two open and two enclosed, extending from a central square platform measuring either 5 or 10 cm. The arms typically range from 25 to 50 cm in length and are approximately 5 cm in width. Open arms are bordered by low side rails (~0.5 cm) to prevent falls while maintaining exposure, whereas enclosed arms are surrounded by high walls, generally 15–40 cm in height, to provide a sense of security. The entire apparatus is elevated 50–70 cm above the floor, and is usually constructed from opaque white or dark-colored plastic to minimize reflections and distractions.

  An overhead video camera system, connected to behavioral tracking software such as EthoVision (Noldus), is commonly integrated for automated data acquisition. Variations of this setup, such as the Elevated Zero Maze, may be used to mitigate the ambiguity associated with the central platform and to further refine anxiety measurements.
- **Testing Procedure**

  To begin the test, the rodent is gently placed at the center of the maze, facing an open or enclosed arm, and allowed to explore freely for a defined period, typically ranging from 5 to 10 minutes. Key behavioral parameters include:
  - Time spent in open versus closed arms
  - Number of entries into each arm
  - Total distance traveled

Rodents naturally avoid open, elevated spaces due to an innate fear of heights and exposure. Therefore, increased time spent in the closed arms, along with reduced entries into open arms, is interpreted as indicative of elevated anxiety levels. Conversely, increased exploration of open arms is generally considered a sign of reduced anxiety or heightened exploratory drive. To maintain experimental rigor and minimize olfactory cues, the maze should be thoroughly cleaned with 70% ethanol between trials (72,73) (Figure 8.14).

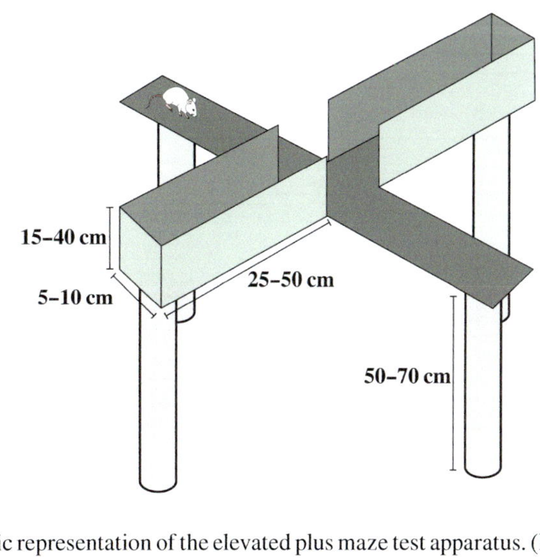

FIGURE 8.14 Schematic representation of the elevated plus maze test apparatus. (Photograph by the authors.)

8.5.5 Marble Burying Test

The Marble burying test (MBT) is a widely utilized behavioral assay in rodents, particularly mice, for modeling and assessing anxiety-like and obsessive–compulsive disorder (OCD)-like behaviors (74). This task exploits the rodent's natural tendency to dig and bury objects in their environment and is considered a sensitive measure of repetitive and compulsive-like behaviors.

- **Apparatus Design and Setup**

  Testing is conducted in a standard polycarbonate rat cage measuring approximately 26 × 48 × 20 cm, filled with a 5 cm-deep layer of unscented mouse bedding. Prior to testing, animals are habituated to the testing environment without the presence of marbles, typically for one day.

  On the test day, 15–20 clean, dry glass marbles (approximately 15 mm in diameter) are evenly spaced on the bedding surface. The test mouse is then gently placed in a corner of the cage, away from the marbles, and the cage is covered with a ventilated filter-top lid. Food and water are withheld during the brief testing period to prevent distractions.
- **Testing Procedure**

  The mouse is allowed to explore the environment and interact with the marbles freely for a 30-minute session. A marble is considered buried when at least two-thirds of its surface area is covered by bedding. Key behavioral parameters include:
  - Latency to initiate burying behavior
  - Total number of marbles buried
  - Digging behavior and bedding displacement (optional, for assessing repetitive motor activity)

Increased marble burying is typically interpreted as an index of heightened anxiety or compulsive-like behavior, although interpretation can vary depending on the pharmacological or genetic model under study. Conversely, reduced burying behavior following anxiolytic or anti-compulsive treatment is often taken as evidence of therapeutic efficacy (74,75) (Figure 8.15).

8.5.6 Social Interaction Tests

8.5.6.1 Reciprocal Social Interaction Test

The social interaction test is a widely used behavioral assay designed to evaluate social behavior and affiliative interactions in rodents, typically between genetically similar or familiar conspecifics,

FIGURE 8.15 Schematic representation of the marble burying test apparatus. (Photograph by the authors.)

under controlled conditions. This test allows for the quantification of both prosocial and avoidant behaviors, and is particularly useful in assessing alterations in social motivation, recognition, and anxiety-related responses in various experimental models.

- **Apparatus Design and Setup**

  For rats, the test arena consists of a floor area measuring 90 × 90 cm, enclosed by 100 cm high transparent Perspex walls to provide a contained but visible environment. Illumination is provided by an infrared lighting system, making the setup suitable for nocturnal animals without disturbing their natural behavior. An overhead infrared camera enables continuous video monitoring and behavioral analysis. To prevent olfactory interference between trials, the arena is thoroughly cleaned with 70% ethanol between sessions.
- **Testing Procedure**

  The testing protocol varies across different studies; we introduce only one of them. It spans five consecutive days and is structured to progressively habituate the animals to the testing environment, minimize novelty-induced responses, and isolate social variables (76).
  - *Day 1 – Habituation 1 (Individual Exposure)*:

    Each animal is placed individually into the test arena for a 20-minute habituation session to familiarize it with the environment.
  - *Day 2 – Habituation 2 (Paired Exposure)*:

    Pairs of cagemate rats are placed together in the arena for 20 minutes to habituate them to the setting under social conditions and reduce novelty effects.
  - *Day 3 – Rest Day*:

    No procedures are performed. This recovery period minimizes stress accumulation and allows consolidation of habituation effects.
  - *Day 4 – Social Isolation*:

    Cagemates are separated and housed individually for 24 hours. During isolation, the animals remain in the same room, allowing auditory and olfactory interaction while preventing visual and physical contact. This mild deprivation enhances motivation for social interaction during the test.
  - *Day 5 – Social Interaction Test*:

    Cagemates are reunited in the test arena for a 20-minute social interaction session.

Behavioral analysis is typically categorized into four domains:

1. Social contact behaviors involve direct physical interaction between conspecifics and provide insight into the quality and intensity of social engagement. These behaviors are further categorized into nape attacking, pinning, allogrooming, and social nose contact.
2. Social interest refers to non-aggressive, non-contact behaviors that reflect an animal's motivation to engage with a conspecific. Two key behaviors commonly scored in this category include:
   - *Approaching*: This behavior is defined by the focal animal moving directly toward another animal, resulting in a clear decrease in distance between the two.
   - *Following*: In this behavior, the focal animal actively maintains close proximity to another animal by moving along with it as it navigates the environment.
3. Social Avoidance reflects an active effort by an animal to reduce or terminate social contact after having been in close proximity.
4. Non-Social Behavior indicates the animal engages in individual activities, such as self-grooming, rearing, or environmental exploration (Figure 8.16).

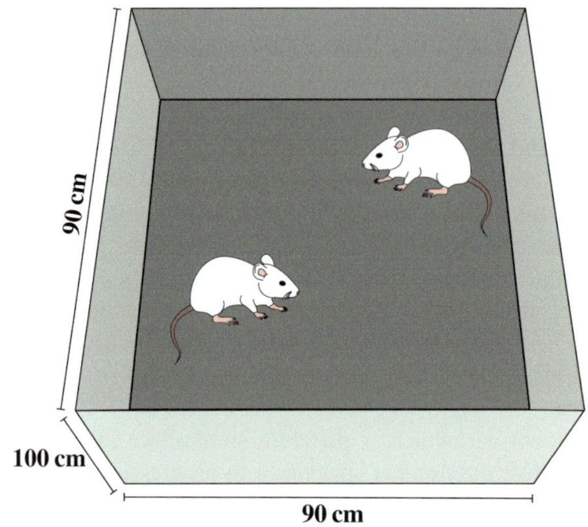

FIGURE 8.16 Schematic representation of the reciprocal social interaction test apparatus. (Photograph by the authors.)

8.5.6.2 Three-Chamber Social Interaction Test

The three-chamber social interaction test is another standard behavioral paradigm used to evaluate sociability and social novelty preference in rodent models.

- **Apparatus Design and Setup**

  Behaviors are assessed in a rectangular black Plexiglas box, which is divided into three interconnected compartments: a central neutral arena and two lateral chambers. The compartments were separated by transparent walls, each with an opening to allow unrestricted access between chambers. Each of the lateral chambers contained a wire mesh enclosure designed to hold stimulus animals. The central compartment remains empty throughout the test. The use of mesh enclosures serves to restrict the movement of the stimulus animals while permitting olfactory, visual, and limited tactile interaction. This restriction is critical to ensure that the social engagement observed during testing is driven primarily by the experimental animal, rather than by the behavior of the stimulus animal. The testing chamber and mesh cages are thoroughly cleaned with 70% ethanol between trials to eliminate residual olfactory cues.
- **Testing Procedure**

  All testing takes place under red light illumination. Animals are habituated to the testing room for at least 1 hour prior to the start of the test. Stimulus animals are naïve, age- and weight-matched wild-type male rats from the same strain as the experimental animals. They are not habituated to the wire mesh cages beforehand.

  In one kind of standard protocol, the procedure consists of three sequential sessions:
  - Habituation Phase

    The experimental rat is placed in the central arena and allowed to explore all three chambers. Both lateral compartments contain empty wire mesh boxes. This phase familiarizes the animal with the environment and controls for any side preference.
  - Social Interaction Phase

    A stranger rat is placed in one of the wire mesh cages in a lateral chamber, while the opposite chamber continues to contain an empty cage. The experimental rat is

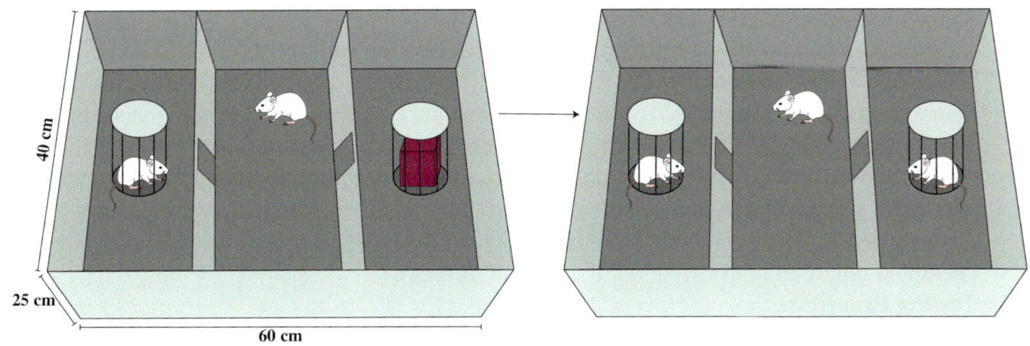

FIGURE 8.17 Schematic representation of the three-chamber social interaction test apparatus. (Photograph by the authors.)

reintroduced into the neutral chamber and allowed to freely explore the entire apparatus. This phase assesses sociability by measuring the animal's preference for the social stimulus (the stranger rat) over the non-social stimulus (the empty cage).

- Social Novelty Phase

  A second, unfamiliar conspecific is introduced into the previously empty mesh cage on the opposite side. The test rat can now choose between the familiar rat encountered in the previous phase and a novel rat. This session evaluates the subject's preference for social novelty.

The key variables measured include:

1. Time spent in each chamber, defined as the test rat entering with at least half of its body
2. Time spent interacting with the wire mesh enclosures, which includes sniffing the box (with visible snout contact), climbing, resting with forelimbs on the cage, and sitting or moving on the cage (Figure 8.17)

These measurements provide quantitative indicators of social approach behavior and social recognition (76).

8.5.7 Motor Function, Balance, and Coordination Tests

8.5.7.1 Rotarod Test

The rotarod test is a widely employed behavioral assay for evaluating motor coordination, balance, grip strength, and motor learning in rodents, particularly in mice and rats (77). Originally developed by Dunham and Miya in 1957 to assess drug-induced motor impairment 704, the test has undergone refinements over time to increase its sensitivity and versatility. In the 1960s, Jones and Roberts introduced an accelerating mode, which gradually increases the rotation speed of the rod. This modification significantly enhances the test's ability to detect subtle motor deficits and learning capacity (78).

Neuroanatomically, rotarod performance engages a distributed neural network comprising the motor cortex, cerebellum, striatum, hippocampus, amygdala, hypothalamus, thalamus, and vestibular nuclei, as well as the associated white matter tracts. These interconnected regions contribute to sensorimotor integration, postural stability, voluntary movement control, emotional regulation during challenge, and the acquisition of motor skills. Studies demonstrate that repeated rotarod training induces both structural and microstructural changes in adult brains, highlighting the test's sensitivity to neuroplasticity (79,80).

- **Apparatus Design and Setup**
  The rotarod apparatus consists of a horizontal rotating rod, typically around 3 cm in diameter for mice, suspended approximately 30 cm above a padded base that prevents injury upon falling. The surface of the rod is often knurled with fine ridges to enhance grip. Two flanges or side barriers are placed along the rod to prevent animals from falling off laterally. Modern rotarod systems are designed to test multiple animals simultaneously, with individual compartments separated by vertical dividers. Most systems are equipped with computerized control units that automatically record key performance metrics, including latency to fall, rotational speed at fall, and trial duration.
- **Testing Procedure**
  The rotarod test is typically conducted in two modes:
  - *Fixed-Speed Mode*: The rod rotates at a constant speed – commonly between 4 and 16 rpm – for a duration of 3–5 minutes. The time the animal remains on the rod is recorded as the primary outcome (Figure 8.18).
  - *Accelerating Mode*: The rotation starts at a low speed (e.g., 4 rpm) and increases steadily – often at a rate of 20 rpm per minute – to a maximum speed ranging from 40 to 80 rpm. This mode is more sensitive in evaluating both motor coordination and endurance.

To minimize experimenter bias, animals are placed on the rod facing away from the researcher. A typical testing protocol begins with a training phase, consisting of several trials at lower speeds to allow animals to acclimate to the task. This is followed by test trials, during which performance is measured under more challenging conditions.

ITIs of 5–10 minutes are implemented to ensure adequate rest and minimize fatigue. Each trial ends either when the animal falls from the rod or reaches the maximum allotted time. Outcomes include both latency to fall and the rotational speed at the time of fall, recorded manually or via automated systems. Repeating trials across multiple days provides a robust measure of motor learning, skill acquisition, and performance consistency (81,82).

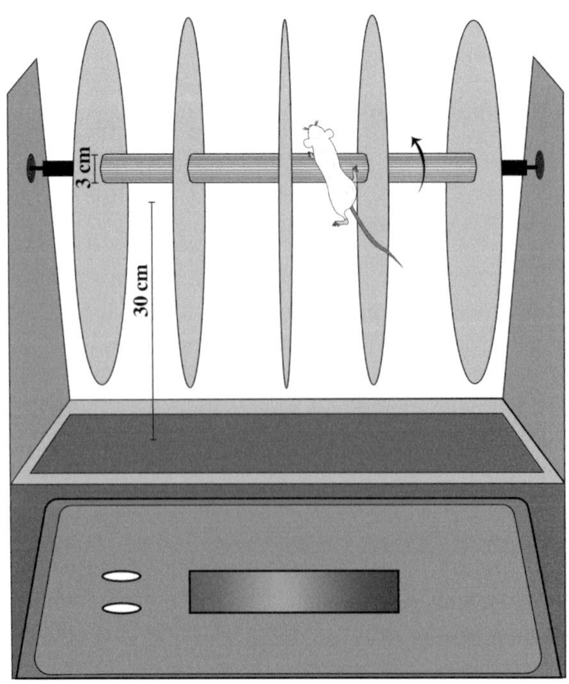

FIGURE 8.18 Schematic representation of the rotarod test apparatus. (Photograph by the authors.)

8.5.7.2 Beam Walk Test

The beam walk test, also known as the balance beam test, is a widely utilized behavioral assay for evaluating motor coordination, balance, and gait abnormalities in rodent models of neurological disorders such as Parkinson's disease. Although commonly applied in animal studies today, the origins of beam-based assessments of balance trace back to at least 1947, when Harold G. Seashore developed a beam-walking test to assess balance development in children (83).

- **Apparatus Design and Setup**
  The standard apparatus consists of a horizontal narrow beam, typically 1 m in length and 1–4 cm in width, elevated 0.5–1 meter above a padded surface to minimize injury in the event of a fall. One end of the beam connects to the animal's home cage or a goal box, which serves as a natural motivator for the rodent to traverse the beam. The opposite end is designated as the starting point.
- **Testing Procedure**
  The beam walk test follows a multi-day protocol typically consisting of two days of training followed by one day of testing. During each training day, animals undergo five trials, with 5-minute ITIs to prevent fatigue and maintain motivation. In each trial, the rodent is gently placed behind the start line and encouraged to cross the beam toward its home cage or goal box.

  On the testing day, the same procedure is followed. The primary measure is the total time on the beam, defined as the duration from placement at the start zone until all four limbs have entered the goal box. A maximum cutoff time of 2 minutes is usually imposed for each trial. Each animal completes five test trials, and the average time across trials is recorded as the final performance score.

  In addition to latency, researchers also record the number of paw slips or foot faults, defined as any instance in which a limb slips off the surface of the beam. Animals with motor impairments, especially those with nigrostriatal dysfunction, typically exhibit longer traversal times and more frequent slips, making these metrics sensitive indicators of sensorimotor deficits (84) (Figure 8.19).

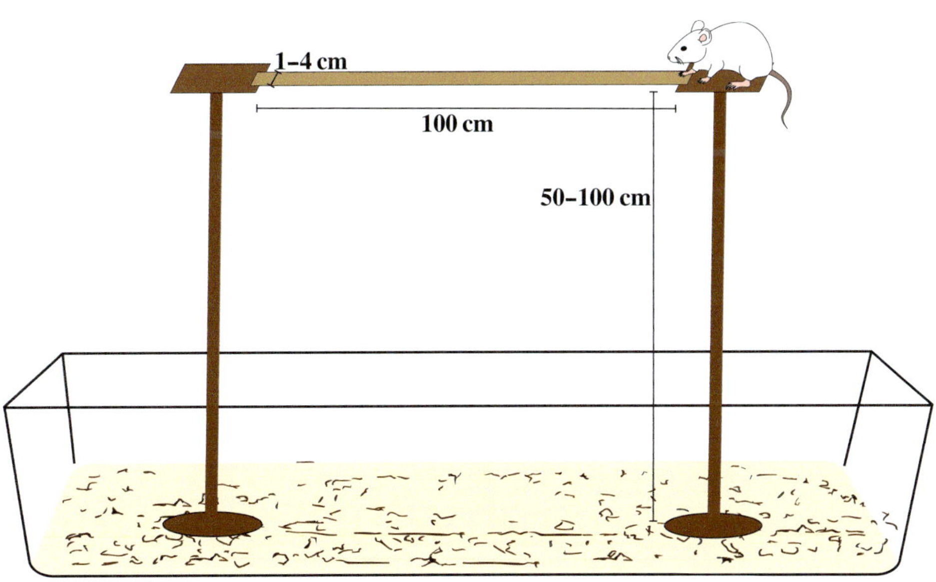

FIGURE 8.19 Schematic representation of the beam walk test apparatus. (Photograph by the authors.)

8.5.7.3 Pole Test

The pole test is a widely applied behavioral assay designed to assess motor dysfunction, particularly bradykinesia and coordination, in rodent models of Parkinson's disease. It is especially sensitive to motor deficits arising from basal ganglia dysfunction, making it a valuable tool in the evaluation of Parkinsonian phenotypes and dopaminergic depletion (85).

- **Apparatus Design and Setup**
  The test apparatus consists of a vertical wooden or metal pole, typically 1 m in height and approximately 3 cm in diameter, mounted upright in the center of an empty cage or designated testing area. The surface of the pole is intentionally smooth, offering minimal grip to allow for sensitive detection of motor impairments such as rigidity and delayed motor responses.
- **Testing Procedure**
  Animals undergo two training sessions prior to the test to habituate them to the task and minimize stress-related variability. During testing, the rodent is gently placed head-upward at the top of the pole, and a stopwatch is started immediately upon placement.
  Two primary parameters are measured:
  - *T-Turn (Turn Time)*: The time required for the animal to reorient itself from a head-up to a head-down position.
  - *Total Descent Time*: The time taken to climb down the entire length of the pole and reach the base or floor of the testing area.

To enhance the reliability of the results, animals typically undergo five trials, with inter-ITIs of a few minutes to prevent fatigue. The mean values of T-turn and total descent times across trials are used for data analysis.

Rodents with intact motor function usually perform the task quickly and smoothly, while those with dopaminergic dysfunction exhibit delayed turning and prolonged descent times, reflecting bradykinesia, akinesia, or postural instability. The pole test is therefore particularly effective for detecting subtle motor deficits and for evaluating the impact of pharmacological or genetic interventions in Parkinson's disease models (Figure 8.20).

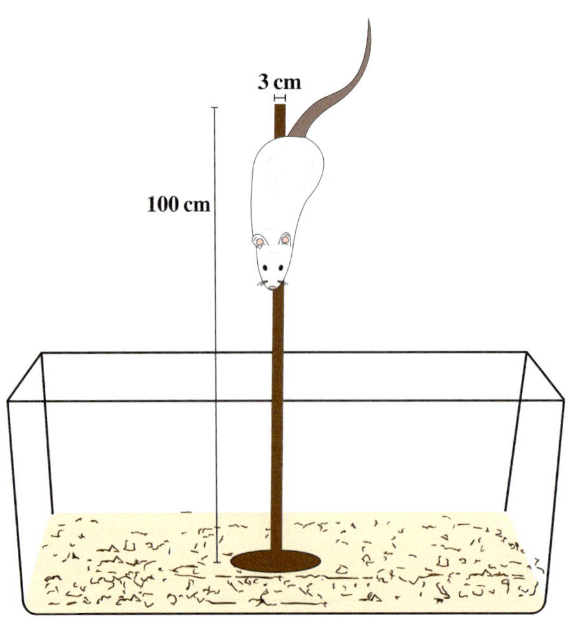

FIGURE 8.20 Schematic representation of the pole test apparatus. (Photograph by the authors.)

8.6 CONCLUSION

Behavioral assays in rodent models provide a powerful toolkit for probing the multifaceted nature of behavior and its underlying neural, genetic, and environmental determinants. The diversity of available paradigms enables researchers to examine specific behavioral domains while also capturing their complex interactions. Continued refinement of these methodologies, alongside the integration of complementary approaches such as neuroimaging, molecular profiling, and computational analysis, will deepen our capacity to link behavioral phenotypes to underlying biological mechanisms. Such advances not only expand our fundamental understanding of brain function but also strengthen the translational bridge to human research, ultimately informing the development of targeted strategies for the prevention, diagnosis, and treatment of neuropsychiatric and neurological disorders.

REFERENCES

1. Snowden JS, Craufurd D, Thompson J, Neary D. Psychomotor, executive, and memory function in preclinical Huntington's disease. Journal of Clinical and Experimental Neuropsychology [Internet]. 2002 [cited 2025 Apr 6];24(2):133–45. Available from: https://pubmed.ncbi.nlm.nih.gov/11992196/
2. Bonifacino T, Zerbo RA, Balbi M, Torazza C, Frumento G, Fedele E, et al. Nearly 30 years of animal models to study amyotrophic lateral sclerosis: A historical overview and future perspectives. International Journal of Molecular Sciences. 2021 Nov;22(22).
3. Sharma S, Rakoczy S, Brown-Borg H. Assessment of spatial memory in mice. Life Sciences. 2010 Oct;87(17–18):521–36.
4. Kim DG, Gonzales EL, Kim S, Kim Y, Adil KJ, Jeon SJ, et al. Social interaction test in home cage as a novel and ethological measure of social behavior in mice. Experimental Neurobiology. 2019 Apr;28(2): 247–60.
5. Morris RGM. Spatial localization does not require the presence of local cues. Learning and Motivation [Internet]. 1981 May 1 [cited 2025 Aug 19];12(2):239–60. Available from: https://www.sciencedirect.com/science/article/abs/pii/0023969081900205
6. Othman MZ, Hassan Z, Che Has AT. Morris water maze: A versatile and pertinent tool for assessing spatial learning and memory. Experimental Animals [Internet]. 2022 [cited 2025 May 31];71(3):264–80. Available from: https://www.jstage.jst.go.jp/article/expanim/71/3/71_21-0120/_article
7. D'Hooge R, De Deyn PP. Applications of the Morris water maze in the study of learning and memory. Brain Research Brain Research Reviews. 2001 Aug;36(1):60–90.
8. Huang Y, Zhou W, Zhang Y. Bright lighting conditions during testing increase thigmotaxis and impair water maze performance in BALB/c mice. Behavioural Brain Research [Internet]. 2012 Jan 1 [cited 2025 Aug 19];226(1):26–31. Available from: https://pubmed.ncbi.nlm.nih.gov/21907245/
9. d'Isa R, Comi G, Leocani L. Apparatus design and behavioural testing protocol for the evaluation of spatial working memory in mice through the spontaneous alternation T-maze. Scientific Reports [Internet]. 2021 Dec 1 [cited 2025 Aug 19];11(1):1–13. Available from: https://www.nature.com/articles/s41598-021-00402-7
10. Rincón-Cortés M, Grace AA. Adaptations in reward-related behaviors and mesolimbic dopamine function during motherhood and the postpartum period. Frontiers in Neuroendocrinology [Internet]. 2020 Apr 1 [cited 2025 Aug 19];57:100839. Available from: https://www.sciencedirect.com/science/article/abs/pii/S0091302220300303
11. Deacon RMJ, Rawlins JNP. T-maze alternation in the rodent. Nature Protocols [Internet]. 2006 Jun 27 [cited 2025 Aug 19];1(1):7–12. Available from: https://www.nature.com/articles/nprot.2006.2
12. Yerkes RM. The intelligence of earthworms. Journal of Animal Behavior [Internet]. 1912 Sep [cited 2025 Jun 7];2(5):332–52. Available from: https://doi.apa.org/doi/10.1037/h0072456
13. Tolman EC. Purpose and cognition: the determiners of animal learning. Psychological Review [Internet]. 1925 Jul [cited 2025 Jun 7];32(4):285–97. Available from: https://doi.apa.org/doi/10.1037/h0072784
14. Dennis W, Henneman RH. The non-random character of initial maze behavior. Pedagogical Seminary and Journal of Genetic Psychology [Internet]. 1932 [cited 2025 Aug 19];40(2):396–405. Available from: https://www.tandfonline.com/doi/abs/10.1080/08856559.1932.10532433

15. Beracochea DJ, Jaffard R. Effects of ibotenic lesions of mammillary bodies on spontaneous and rewarded spatial alternation in mice. Journal of Cognitive Neuroscience [Internet]. 1990 Apr 1 [cited 2025 Jun 7];2(2):133–40. Available from: https://dx.doi.org/10.1162/jocn.1990.2.2.133
16. Thomas GJ. Comparison of effects of small lesions in posterodorsal septum on spontaneous and rerun correction (contingently reinforced) alternation in rats. Journal of Comparative and Physiological Psychology [Internet]. 1979 Aug [cited 2025 Jun 7];93(4):685–94. Available from: https://pubmed.ncbi.nlm.nih.gov/479403/
17. Gerlai R. A new continuous alternation task in T-maze detects hippocampal dysfunction in mice: A strain comparison and lesion study. Behavioural Brain Research [Internet]. 1998 Sep [cited 2025 Jun 7]; 95(1):91–101. Available from: https://pubmed.ncbi.nlm.nih.gov/9754881/
18. Spowart-Manning L, Van Der Staay FJ. The T-maze continuous alternation task for assessing the effects of putative cognition enhancers in the mouse. Behavioural Brain Research [Internet]. 2004 May 5 [cited 2025 Jun 7];151(1–2):37–46. Available from: https://pubmed.ncbi.nlm.nih.gov/15084419/
19. Lalonde R, Strazielle C. Relations between open-field, elevated plus-maze, and emergence tests as displayed by C57/BL6J and BALB/c mice. Journal of Neuroscience Methods [Internet]. 2008 Jun 15 [cited 2025 Jun 7];171(1):48–52. Available from: https://pubmed.ncbi.nlm.nih.gov/18358538/
20. Zhang Q, Kobayashi Y, Goto H, Itohara S. An automated T-maze based apparatus and protocol for analyzing delay- and effort-based decision making in free moving rodents. Journal of Visualized Experiments : JoVE [Internet]. 2018 Aug 2 [cited 2025 Jun 7];2018(138):57895. Available from: https://pmc.ncbi.nlm.nih.gov/articles/PMC6126603/
21. Prieur EAK, Jadavji NM. Assessing spatial working memory using the spontaneous alternation Y-maze test in aged male mice. Bio-Protocol [Internet]. 2019 Feb 5 [cited 2025 Jun 7];9(3). Available from: https://pubmed.ncbi.nlm.nih.gov/33654968/
22. Ghafouri S, Fathollahi Y, Javan M, Shojaei A, Asgari A, Mirnajafi-Zadeh J. Effect of low frequency stimulation on impaired spontaneous alternation behavior of kindled rats in Y-maze test. Epilepsy Research [Internet]. 2016 Oct 1 [cited 2025 Jun 7];126:37–44. Available from: https://pubmed.ncbi.nlm.nih.gov/27423017/
23. Kim J, Kang H, Lee YB, Lee B, Lee D. A quantitative analysis of spontaneous alternation behaviors on a Y-maze reveals adverse effects of acute social isolation on spatial working memory. Scientific Reports 2023 13:1 [Internet]. 2023 Sep 7 [cited 2025 Jun 7];13(1):1–13. Available from: https://www.nature.com/articles/s41598-023-41996-4
24. Kraeuter AK, Guest PC, Sarnyai Z. The Y-Maze for assessment of spatial working and reference memory in mice. Methods in Molecular Biology [Internet]. 2019 [cited 2025 Jun 7];1916:105–11. Available from: https://pubmed.ncbi.nlm.nih.gov/30535688/
25. Barnes CA. Memory deficits associated with senescence: A neurophysiological and behavioral study in the rat. Journal of Comparative and Physiological Psychology [Internet]. 1979 Feb [cited 2025 Jun 7];93(1):74–104. Available from: https://pubmed.ncbi.nlm.nih.gov/221551/
26. Pitts M. Barnes maze procedure for spatial learning and memory in mice. Bio-Protocol [Internet]. 2018 [cited 2025 Jun 7];8(5). Available from: https://pubmed.ncbi.nlm.nih.gov/29651452/
27. Rodríguez Peris L, Scheuber MI, Shan H, Braun M, Schwab ME. Barnes maze test for spatial memory: A new, sensitive scoring system for mouse search strategies. Behavioural Brain Research [Internet]. 2024 Feb 26 [cited 2025 Jun 7];458. Available from: https://pubmed.ncbi.nlm.nih.gov/37898351/
28. Rosenfeld CS, Ferguson SA. Barnes maze testing strategies with small and large rodent models. Journal of Visualized Experiments [Internet]. 2014 Feb 26 [cited 2025 Jun 7];(84). Available from: https://pubmed.ncbi.nlm.nih.gov/24637673/
29. Vogel-Ciernia A, Wood MA. Examining object location and object recognition memory in mice. Current Protocols in Neuroscience [Internet]. 2014 [cited 2025 Jun 27];2014:8.31.1-8.31.17. Available from: https://pubmed.ncbi.nlm.nih.gov/25297693/
30. Ammassari-Teule M, Passino E. The dorsal hippocampus is selectively involved in the processing of spatial information even in mice with a genetic hippocampal dysfunction. Psychobiology [Internet]. 1997 Jun 4 [cited 2025 Jun 27];25(2):118–25. Available from: https://link.springer.com/article/10.3758/BF03331916
31. Bevins RA, Besheer J, Palmatier MI, Jensen HC, Pickett KS, Eurek S. Novel-object place conditioning: behavioral and dopaminergic processes in expression of novelty reward. Behavioural Brain Research [Internet]. 2002 Feb 1 [cited 2025 Jun 27];129(1–2):41–50. Available from: https://www.sciencedirect.com/science/article/abs/pii/S0166432801003266

32. Goulart BK, de Lima MNM, de Farias CB, Reolon GK, Almeida VR, Quevedo J, et al. Ketamine impairs recognition memory consolidation and prevents learning-induced increase in hippocampal brain-derived neurotrophic factor levels. Neuroscience [Internet]. 2010 Jun [cited 2025 Jun 27];167(4):969–73. Available from: https://pubmed.ncbi.nlm.nih.gov/20338225/
33. Silvers JM, Harrod SB, Mactutus CF, Booze RM. Automation of the novel object recognition task for use in adolescent rats. Journal of Neuroscience Methods [Internet]. 2007 Oct 15 [cited 2025 Jun 27];166(1):99–103. Available from: https://pubmed.ncbi.nlm.nih.gov/17719091/
34. Antunes M, Biala G. The novel object recognition memory: Neurobiology, test procedure, and its modifications. Cognitive Processing [Internet]. 2012 May [cited 2025 Jun 27];13(2):93–110. Available from: https://pubmed.ncbi.nlm.nih.gov/22160349/
35. Ennaceur A, Delacour J. A new one-trial test for neurobiological studies of memory in rats. 1: Behavioral data. Behavioural Brain Research [Internet]. 1988 Nov 1 [cited 2025 Jun 27];31(1):47–59. Available from: https://pubmed.ncbi.nlm.nih.gov/3228475/
36. Hammond C. Impacts of lifelong learning upon emotional resilience, psychological and mental health: Fieldwork evidence. Oxford Review of Education [Internet]. 2004 Dec [cited 2025 Jun 27];30(4):551–68. Available from: https://www.tandfonline.com/doi/abs/10.1080/0305498042000303008
37. Mumby DG, Gaskin S, Glenn MJ, Schramek TE, Lehmann H. Hippocampal damage and exploratory preferences in rats: Memory for objects, places, and contexts. Learning and Memory [Internet]. 2002 [cited 2025 Jun 27];9(2):49–57. Available from: https://pubmed.ncbi.nlm.nih.gov/11992015/
38. Koivisto H, Maguire C, Tanila H. Novel object recognition task for mice: Is it a test for memory, object neophobia or innate preference? Behavioural Brain Research [Internet]. 2025 Aug 5 [cited 2025 Jun 2]; 491:115649. Available from: https://linkinghub.elsevier.com/retrieve/pii/S0166432825002359
39. Kalueff AV, Keisala T, Minasyan A, Kuuslahti M, Tuohimaa P. Temporal stability of novelty exploration in mice exposed to different open field tests. Behavioural Processes [Internet]. 2006 Mar [cited 2025 Jun 3];72(1):104–12. Available from: https://linkinghub.elsevier.com/retrieve/pii/S037663570500272X
40. Himanshu K, Dharmila, Sarkar D, Nutan. A review of behavioral tests to evaluate different types of anxiety and anti-anxiety effects. Clinical Psychopharmacology and Neuroscience [Internet]. 2020 May 1 [cited 2025 Jun 8];18(3):341–51. Available from: https://pubmed.ncbi.nlm.nih.gov/32702213/
41. Hall CS. Emotional behavior in the rat. I. Defecation and urination as measures of individual differences in emotionality. Journal of Comparative Psychology. 1934 Dec;18(3):385–403.
42. Gencturk S, Unal G. Rodent tests of depression and anxiety: Construct validity and translational relevance. Cognitive, Affective and Behavioral Neuroscience [Internet]. 2024 Apr 1 [cited 2025 Jun 8]; 24(2):191–224. Available from: https://pubmed.ncbi.nlm.nih.gov/38413466/
43. Kraeuter AK, Guest PC, Sarnyai Z. The open field test for measuring locomotor activity and anxiety-like behavior. Methods in Molecular Biology [Internet]. 2019 [cited 2025 Jun 8];1916:99–103. Available from: https://pubmed.ncbi.nlm.nih.gov/30535687/
44. Takeda H, Tsuji M, Matsumiya T. Changes in head-dipping behavior in the hole-board test reflect the anxiogenic and/or anxiolytic state in mice. European Journal of Pharmacology [Internet]. 1998 May 29 [cited 2025 Jun 8];350(1):21–9. Available from: https://pubmed.ncbi.nlm.nih.gov/9683010/
45. Brown GR, Nemes C. The exploratory behaviour of rats in the hole-board apparatus: Is head-dipping a valid measure of neophilia? Behavioural Processes [Internet]. 2008 Jul 1 [cited 2025 Jun 8];78(3):442–8. Available from: https://www.sciencedirect.com/science/article/pii/S0376635708000983
46. Casarrubea M, Di Giovanni G, Aiello S, Crescimanno G. The hole-board apparatus in the study of anxiety. Physiology and Behavior [Internet]. 2023 Nov 1 [cited 2025 Jun 27];271. Available from: https://pubmed.ncbi.nlm.nih.gov/37690695/
47. Shoji H, Takao K, Hattori S, Miyakawa T. Contextual and cued fear conditioning test using a video analyzing system in mice. Journal of Visualized Experiments [Internet]. 2014 Mar 1 [cited 2025 Jun 7];(85). Available from: https://pubmed.ncbi.nlm.nih.gov/24637495/
48. Curzon P, Rustay NR, Browman KE. Cued and Contextual Fear Conditioning for Rodents. 2009 Oct 29 [cited 2025 Jun 8];19–37. Available from: https://www.ncbi.nlm.nih.gov/books/NBK5223/
49. Atallah HE, Frank MJ, O'Reilly RC. Hippocampus, cortex, and basal ganglia: Insights from computational models of complementary learning systems. Neurobiology of Learning and Memory [Internet]. 2004 Nov [cited 2025 Jun 8];82(3):253–67. Available from: https://pubmed.ncbi.nlm.nih.gov/15464408/
50. Anagnostaras SG, Gale GD, Fanselow MS. Hippocampus and contextual fear conditioning: Recent controversies and advances. Hippocampus [Internet]. 2001 [cited 2025 Jun 8];11(1):8–17. Available from: https://pubmed.ncbi.nlm.nih.gov/11261775/

51. Sanei M, Saberi-Demneh A. Effect of curcumin on memory impairment: A systematic review. Phytomedicine [Internet]. 2019 Jan 1 [cited 2025 Jun 8];52:98–106. Available from: https://pubmed.ncbi.nlm.nih.gov/30599917/
52. Filho GLB, Zenki KC, Kalinine E, Baggio S, Pettenuzzo L, Zimmer ER, et al. A new device for step-down inhibitory avoidance task – Effects of low and high frequency in a novel device for passive inhibitory avoidance task that avoids bioimpedance variations. PLoS One [Internet]. 2015 Feb 23 [cited 2025 Jun 8];10(2). Available from: https://pubmed.ncbi.nlm.nih.gov/25706879/
53. Taherianfard M, Ahmadijokani S. The passive avoidance task ameliorate the toxic effects of bisphenol A on dopamine D1 receptor density in hippocampus, amygdala, and cerebellum of male rats. Brain and Behavior [Internet]. 2023 Apr 1 [cited 2025 Jun 8];13(4). Available from: https://pubmed.ncbi.nlm.nih.gov/36879399/
54. Lénárd L, Kertes E. Influence of passive avoidance learning by substance P in the basolateral amygdala. Acta Biologica Hungarica [Internet]. 2002 [cited 2025 Jun 8];53(1–2):95–104. Available from: https://pubmed.ncbi.nlm.nih.gov/12064784/
55. Jänicke B, Coper H. Tests in rodents for assessing sensorimotor performance during aging. Advances in Psychology [Internet]. 1996 Jan 1 [cited 2025 Jun 8];114(C):201–33. Available from: https://www.sciencedirect.com/science/article/abs/pii/S0166411596800100
56. Eagle A, Wang H, Robison A. Sensitive assessment of hippocampal learning using temporally dissociated passive avoidance task. Bio-Protocol [Internet]. 2016 [cited 2025 Jun 8];6(11). Available from: https://pubmed.ncbi.nlm.nih.gov/29119127/
57. Ghafarimoghadam M, Mashayekh R, Gholami M, Fereydani P, Shelley-Tremblay J, Kandezi N, et al. A review of behavioral methods for the evaluation of cognitive performance in animal models: Current techniques and links to human cognition. Physiology and Behavior [Internet]. 2022 Feb 1 [cited 2025 Jun 8];244. Available from: https://pubmed.ncbi.nlm.nih.gov/34801559/
58. Cline BH, Steinbusch HWM, Malin D, Revishchin AV, Pavlova GV, Cespuglio R, et al. The neuronal insulin sensitizer dicholine succinate reduces stress-induced depressive traits and memory deficit: Possible role of insulin-like growth factor 2. BMC Neuroscience [Internet]. 2012 Sep 18 [cited 2025 Jun 8]; 13(1). Available from: https://pubmed.ncbi.nlm.nih.gov/22989159/
59. Garín-Aguilar ME, Medina AC, Quirarte GL, Mcgaugh JL, Prado-Alcalá RA. Intense aversive training protects memory from the amnestic effects of hippocampal inactivation. Hippocampus [Internet]. 2014 Jan [cited 2025 Jun 8];24(1):102–12. Available from: https://pubmed.ncbi.nlm.nih.gov/24123595/
60. Pandya M, Altinay M, Malone DA, Anand A. Where in the brain is depression? Current Psychiatry Reports [Internet]. 2012 Dec [cited 2025 Jun 8];14(6):634–42. Available from: https://pubmed.ncbi.nlm.nih.gov/23055003/
61. Maletic V, Robinson M, Oakes T, Iyengar S, Ball SG, Russell J. Neurobiology of depression: An integrated view of key findings. International Journal of Clinical Practice [Internet]. 2007 Dec [cited 2025 Jun 8];61(12):2030–40. Available from: https://pubmed.ncbi.nlm.nih.gov/17944926/
62. Can A, Dao DT, Arad M, Terrillion CE, Piantadosi SC, Gould TD. The mouse forced swim test. Journal of Visualized Experiments [Internet]. 2011 [cited 2025 Jun 8];(58). Available from: https://pubmed.ncbi.nlm.nih.gov/22314943/
63. Joodaki M, Hosseini N, Joodaki M, Hosseini N, Joodaki M, Hosseini N. Direct and indirect factors affecting the forced swim test to investigate the level of depression in rodents. Thrita Journal of Neuron 2023 12:2 [Internet]. 2023 Mar 5 [cited 2025 Jun 8];12(2). Available from: https://brieflands.com/articles/thrita-144117
64. Yankelevitch-Yahav R, Franko M, Huly A, Doron R. The forced swim test as a model of depressive-like behavior. Journal of Visualized Experiments [Internet]. 2015 Mar 2 [cited 2025 Jun 8];2015(97). Available from: https://pubmed.ncbi.nlm.nih.gov/25867960/
65. de Oliveira-Júnior BA, Marques DB, Rossignoli MT, Prizon T, Leite JP, Ruggiero RN. Multidimensional behavioral profiles associated with resilience and susceptibility after inescapable stress. Scientific Reports [Internet]. 2024 Dec 1 [cited 2025 Jun 8];14(1). Available from: https://pubmed.ncbi.nlm.nih.gov/38678053/
66. Can A, Dao DT, Terrillion CE, Piantadosi SC, Bhat S, Gould TD. The tail suspension test. Journal of Visualized Experiments [Internet]. 2011 [cited 2025 Jun 8];(58). Available from: https://pubmed.ncbi.nlm.nih.gov/22315011/
67. Cryan JF, Mombereau C, Vassout A. The tail suspension test as a model for assessing antidepressant activity: Review of pharmacological and genetic studies in mice. Neuroscience and Biobehavioral Reviews [Internet]. 2005 [cited 2025 Jun 8];29(4–5):571–625. Available from: https://pubmed.ncbi.nlm.nih.gov/15890404/

68. O'Leary OF, Cryan JF. The tail-suspension test: A model for characterizing antidepressant activity in mice. In *Mood and anxiety related phenotypes in mice: characterization using behavioral tests* (pp. 119–137). Totowa, NJ: Humana Press, 2009 [cited 2025 Jun 8]. Available from: https://link.springer.com/protocol/10.1007/978-1-60761-303-9_7
69. Primo MJ, Fonseca-Rodrigues D, Almeida A, Teixeira PM, Pinto-Ribeiro F. Sucrose preference test: A systematic review of protocols for the assessment of anhedonia in rodents. European Neuropsychopharmacology [Internet]. 2023 Dec 1 [cited 2025 Jun 8];77:80–92. Available from: https://pubmed.ncbi.nlm.nih.gov/37741164/
70. Liu MY, Yin CY, Zhu LJ, Zhu XH, Xu C, Luo CX, et al. Sucrose preference test for measurement of stress-induced anhedonia in mice. Nature Protocols [Internet]. 2018 Jul 1 [cited 2025 Jun 8];13(7): 1686–98. Available from: https://pubmed.ncbi.nlm.nih.gov/29988104/
71. Salazar A, Gonzalez-Rivera BL, Redus L, Parrott JM, O'Connor JC. Indoleamine 2,3-dioxygenase mediates anhedonia and anxiety-like behaviors caused by peripheral lipopolysaccharide immune challenge. Hormones and Behavior [Internet]. 2012 Aug [cited 2025 Jun 8];62(3):202–9. Available from: https://pubmed.ncbi.nlm.nih.gov/22504306/
72. Komada M, Takao K, Miyakawa T. Elevated plus maze for mice. Journal of Visualized Experiments [Internet]. 2008 [cited 2025 Jun 15];(22). Available from: https://pubmed.ncbi.nlm.nih.gov/19229173/
73. Ghassemi S, Pourbadie HG, Prehaud C, Lafon M, Sayyah M. Role of the glycoprotein thorns in anxious effects of rabies virus: Evidence from an animal study. Brain Research Bulletin [Internet]. 2022 Jul 1 [cited 2025 Jun 15];185:107–16. Available from: https://pubmed.ncbi.nlm.nih.gov/35537567/
74. Walf AA, Frye CA. The use of the elevated plus maze as an assay of anxiety-related behavior in rodents. Nature Protocols [Internet]. 2007 Mar [cited 2025 Jun 15];2(2):322–8. Available from: https://pubmed.ncbi.nlm.nih.gov/17406592/
75. Angoa-Pérez M, Kane MJ, Briggs DI, Francescutti DM, Kuhn DM. Marble burying and nestlet shredding as tests of repetitive, compulsive-like behaviors in mice. Journal of Visualized Experiments : JoVE [Internet]. 2013 [cited 2025 Jun 15];(82). Available from: https://pubmed.ncbi.nlm.nih.gov/24429507/
76. Manfré G, Novati A, Faccini I, Rossetti AC, Bosch K, Molteni R, et al. BACHD rats expressing full-length mutant huntingtin exhibit differences in social behavior compared to wild-type littermates. PLoS One [Internet]. 2018 Feb 1 [cited 2025 Jun 15];13(2):e0192289. Available from: https://journals.plos.org/plosone/article?id=10.1371/journal.pone.0192289
77. Mouzon B, Chaytow H, Crynen G, Bachmeier C, Stewart J, Mullan M, et al. Repetitive mild traumatic brain injury in a mouse model produces learning and memory deficits accompanied by histological changes. Journal of Neurotrauma [Internet]. 2012 Dec 10 [cited 2025 Jun 27];29(18):2761–73. Available from: https://pubmed.ncbi.nlm.nih.gov/22900595/
78. Jones BJ, Roberts DJ. The quantitative measurement of motor inco-ordination in naive mice using an accelerating rotarod. Journal of Pharmacy and Pharmacology [Internet]. 1968 [cited 2025 Jun 27]; 20(4):302–4. Available from: https://pubmed.ncbi.nlm.nih.gov/4384609/
79. Cording KR, Bateup HS. Altered motor learning and coordination in mouse models of autism spectrum disorder. Frontiers in Cellular Neuroscience [Internet]. 2023 [cited 2025 Jun 27];17. Available from: https://pubmed.ncbi.nlm.nih.gov/38026686/
80. Scholz J, Niibori Y, Frankland W P, Lerch J P. Rotarod training in mice is associated with changes in brain structure observable with multimodal MRI. NeuroImage [Internet]. 2015 Feb 5 [cited 2025 Jun 27]; 107:182–9. Available from: https://pubmed.ncbi.nlm.nih.gov/25497397/
81. Deacon RMJ. Measuring motor coordination in mice. Journal of Visualized Experiments : JoVE [Internet]. 2013 [cited 2025 Jun 27];(75). Available from: https://pubmed.ncbi.nlm.nih.gov/23748408/
82. Nagai H, De Vivo L, Bellesi M, Ghilardi MF, Tononi G, Cirelli C. Sleep consolidates motor learning of complex movement sequences in mice. Sleep [Internet]. 2017 Feb 1 [cited 2025 Jun 27];40(2). Available from: https://pubmed.ncbi.nlm.nih.gov/28364506/
83. Sheta R, Bérard M, Musiol D, Martínez-Drudis L, Oueslati A. Behavioral analysis of motor and non-motor impairment in rodent models of Parkinson's disease. Frontiers in Aging Neuroscience [Internet]. 2024 Dec 23 [cited 2025 Jun 27];16. Available from: https://pubmed.ncbi.nlm.nih.gov/39763579/
84. Luong TN, Carlisle HJ, Southwell A, Patterson PH. Assessment of motor balance and coordination in mice using the balance beam. Journal of Visualized Experiments [Internet]. 2011 [cited 2025 Jun 27]; (49). Available from: https://pubmed.ncbi.nlm.nih.gov/21445033/
85. Glajch KE, Fleming SM, Surmeier DJ, Osten P. Sensorimotor assessment of the unilateral 6-hydroxydopamine mouse model of Parkinson's disease. Behavioural Brain Research [Internet]. 2012 May 1 [cited 2025 Jun 27];230(2):309–16. Available from: https://pubmed.ncbi.nlm.nih.gov/22178078/

Index

Note: Page numbers in **bold** and *italics* refer to tables and figures, respectively.

S

T

U

V

W

X

Y

Z